Geology of the country around Harlech

The area described in this memoir lies almost entirely within the Snowdonia National Park. Spectacular mountainous scenery includes the Harlech Dome, a complex structure of Cambrian sedimentary rocks bounded on the east and south-east by volcanic rocks mainly of Ordovician age. There are many intrusions related to the major volcanic episodes. The area contains mineral deposits of various kinds, including the veins of the Dolgellau gold-belt, and relics of the old mining industries are common.

In this work the stratigraphy and palaeontology of the sedimentary succession and the history of volcanism are described. In addition the petrology and geochemistry of the intrusive rocks is treated in some detail. New interpretations on metallogenesis are described and discussed, and a full account is given of the geophysical work that has been carried out over the Harlech Dome.

Offshore and under the coastal sand-dunes of Morfa Dyffryn lies a thick sequence of Triassic, early Jurassic and Oligocene sediments below Pleistocene and Recent deposits. They were proved by deep drilling which added a new dimension to the geology of the region.

The Rhinogs: Diffwys to Rhinog Fawr (right) forming the western limb of the Dolwen Pericline. *Frontispiece*

BRITISH GEOLOGICAL SURVEY

P. M. ALLEN and
A. A. JACKSON

Geology of the country around Harlech

CONTRIBUTORS
P. N. Dunkley,
H. C. Ivimey-Cook,
B. P. Kokelaar,
W. J. Rea,
C. C. Rundle,
A. W. A. Rushton,
T. J. Shepherd and
I. F. Smith

Memoir for 1:50 000 geological sheet 135 with part of sheet 149, (England and Wales)

1835 Geological Survey of Great Britain
150 Years of Service to the Nation
1985 British Geological Survey

Natural Environment Research Council

LONDON HER MAJESTY'S STATIONERY OFFICE 1985

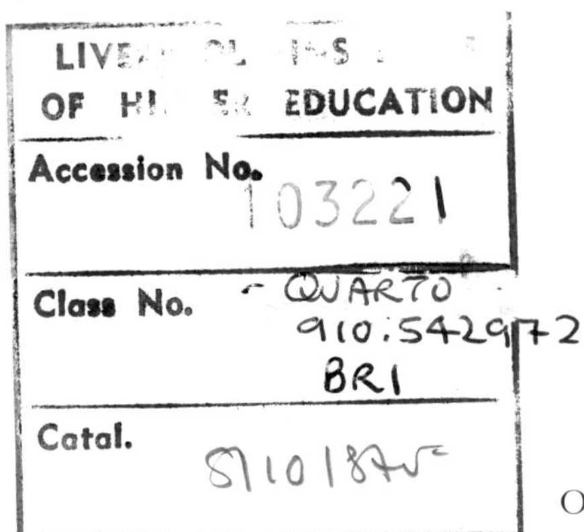

First published 1985

ISBN 0 11 884367 2

Bibliographical reference

Allen, P. M. and Jackson, A. A. 1985.
Geology of the country around Harlech *Mem. Br. Geol. Surv.*, Sheet 135 with part of 149, 112 pp.

Authors

P. M. Allen, BSc, PhD
British Geological Survey,
Windsor Court, Windsor Terrace, Newcastle upon Tyne NE2 4HE

Audrey A. Jackson, BA, PhD
British Geological Survey,
Keyworth, Nottingham NG12 5GG

Contributors

P. N. Dunkley, BSc, PhD
H. C. Ivimey-Cook, BSc, PhD
A. W. A. Rushton, BA, PhD, and
I. F. Smith, BSc, MSc
British Geological Survey, Keyworth

C. C. Rundle, BSc, and
T. J. Shepherd, BSc, PhD
British Geological Survey,
64 Gray's Inn Road, London WC1X 8NG

B. P. Kokelaar, BSc, MSc, PhD
Department of Environmental Studies,
University of Ulster,
Jordanstown,
Newtownabbey,
Co. Antrim BT 37 0QB

W. J. Rea, BA, PhD
Department of Geology and Physical Sciences,
Oxford Polytechnic,
Headington, OXON OX3 0BP

Other publications of the Survey dealing with this and nearby districts

Books

British Regional Geology
North Wales, 3rd Edition, 1961

Memoir and Sheet Explanation
Bangor (106), 1985
Rhyl and Denbigh, (107), 1984

Classical areas of British Geology
Geological excursions in the Harlech Dome, 1985

Maps

1:625 000
Sheet 2 Geological
Sheet 2 Quaternary

1:250 000
Cardigan Bay (52N 06W) Geology 1982
Liverpool Bay (53N 04W) Geology 1978
Sea bed sediments and Quaternary 1984
Anglesey (53N 06W) Geology 1981

1:50 000
Sheet 106 (Bangor) 1985
Sheet 107 (Denbigh) 1973
Sheet 136 (Bala) *in press*
Sheet 163 (Aberystwyth) 1984

CONTENTS

PLATES

FIGURES

TABLES

PREFACE

This memoir describes the geology of the district covered by 1:50 000 Sheet 135 and a small portion of Sheet 149. The primary survey at one-inch to one mile of Sheet 75 SE which is approximately the area of Sheet 135, was carried out by A. C. Ramsay, A. R. C. Selwyn and W. T. Aveline and published in 1855. No sheet memoir, however, was written, the findings of the surveyors being incorporated in Ramsay's memoir on the geology of the whole of North Wales published first in 1866 and in a second edition in 1881.

Reports of gold, found in veins already being mined for copper, lead and silver, came in 1844, shortly before the Geological Survey began work in North Wales, and throughout the middle 19th century officers of the Geological Survey contributed to research on the gold field. Manganese was discovered and mined shortly after gold, and these two metals, together with copper, provided the economic focus in the area until after the First World War. The Special Reports on the Mineral Resources on the metalliferous deposits in this area published by the Geological Survey mostly in the 1920s came largely when mining had ended, and during the next forty years interest in the economic geology of the Harlech Dome declined. During this period, however, A. K. Wells published his work on Rhobell Fawr, and C. A. Matley and T. S. Wilson finished their classic account of the Cambrian of the Harlech Dome, fulfilling Charles Lapworth's expectation that the stratigraphy could be resolved in detail.

Interest in the economic geology of North Wales as a whole revived in the mid-1960s, and in the Harlech Dome it was revitalised in about 1970 when Riofinex Ltd discovered porphyry copper mineralisation in an intrusion complex believed to be genetically associated with the late Tremadoc Rhobell Fawr magmatism. At this time the decision was taken by the Geological Survey to prepare the 1:50 000 Sheet 135 as a provisional map, using line-work taken from field maps of Matley and Wilson for the area they covered and remapping the area east of their ground to the sheet margin at a scale of 6 inches to one mile (1:10 560). The work of remapping solid and drift in the eastern area was begun in 1972 by P. M. Allen and Audrey A. Jackson. In 1972 and 1973 two research students, B. P. Kokelaar and P. N. Dunkley of the University College of Wales, Aberystwyth, began field work on Rhobell Fawr and the south-west Aran mountains respectively. Their work has been included in the remapped area. When the eastern area was complete the drift over Matley and Wilson's ground was mapped at 1:25 000 in 1974–75 by Jackson and Allen. As part of the stratigraphic investigation the Bryn-teg Borehole was drilled at the centre of the Dolwen Pericline in 1972. The Llanbedr (Mochras Farm) Borehole, which had a profound effect on the interpretation of the geology of Cardigan Bay, had already been drilled in 1967–69.

Soon after mapping began the eastern area was covered by an airborne geophysical survey at the start of a mineral exploration project under the Mineral Reconnaissance Programme carried out by the Geological Survey on behalf of the Department of Industry. Throughout the field programme mapping and mineral exploration were carried out side by side.

The memoir has mostly been written by Allen and Jackson. It was compiled by Allen and edited by R. A. B. Bazley. However, it has been very much a collaborative effort including Kokelaar, Ulster Polytechnic, Newtonabbey, who wrote most of the section on the Rhobell Volcanic Group, and W. J. Rea, Oxford Polytechnic, who undertook the geochemistry and contributed to the petrography of the intrusive rocks. A. W. A. Rushton, ably helped in the field by S. P. Tunnicliff, is responsible for the palaeontology. I. F. Smith, who carried out most of the geophysical field investigations under the Mineral Reconnaissance Programme, wrote the chapter on geophysical investigations. Other contributions were made by Dunkley on the Aran Volcanic Group and intrusive rocks, C. C. Rundle on age determinations and T. J. Shepherd on fluid inclusions. H. C. Ivimey-Cook wrote the chapter on the Mesozoic and Tertiary rocks.

Geologists who have contributed indirectly to this memoir, mostly through their work on the Mineral Reconnaissance Programme, include I. R. Basham, D. C. Cooper, G. D. Easterbrook and R. J. Tappin. In the geophysics chapter Smith was able to draw upon the unpublished work of Z. K. Dabek of the Regional Geophysics Research Group, who carried out a gravity survey across the Mochras Fault, and Mrs J. M. Allsop. K. S. Siddiqui carried out X-ray diffraction examinations; K. E. Thornton took the photographs; N. G. Berridge was consulted about petrography.

I am indebted to our collaborators from university departments, and also gratefully acknowledge the invaluable co-operation of local property and mine owners in allowing my staff access to their land and records.

G. M. Brown
Director

British Geological Survey
Keyworth
Nottingham NG12 5GG
10 July 1985

SIX-INCH AND 1:25 000 MAPS

The following list shows the six-inch and 1:25 000 maps included partly or wholly within the area of Sheet 135 (Harlech) of the Geological Map of England and Wales. All the maps are on National Grid lines, lying within the 100 kilometre square SH. Uncoloured dye-line copies of all the maps are available from the Geological Survey in Keyworth. The surveyors of the six-inch maps are P. M. Allen, Audrey A. Jackson, B. P. Kokelaar and P. N. Dunkley. The 1:25 000 maps are compilations of the solid geology by C. A. Matley and T. S. Wilson with a limited amount of remapping and drift lines added by Audrey A. Jackson and P. M. Allen.

Six-inch maps

SH 71 NE	Allen, Jackson, Dunkley	1973–75
SH 71 NW	Allen	1973–74
SH 72 SE	Jackson, Dunkley, Kokelaar	1973–75
SH 72 SW	Allen, Jackson, Kokelaar	1974–75
SH 72 NE	Allen, Jackson, Kokelaar	1973–75
SH 73 SE	Allen, Jackson	1972
SH 73 NE	Allen	1973
SH 81 NW	Dunkley	1973–75
SH 82 SW	Jackson, Dunkley, Allen	1973–75
SH 82 NW	Jackson	1973
SH 83 SW	Allen, Jackson	1973
SH 83 NW	Allen	1973

1:25 000 maps

SH 52	Drift, Jackson	1975
SH 53	Drift, Jackson	1975
	Local solid remapping, Jackson	1974–75
SH 61	Drift, Allen	1975
	Local solid remapping, Allen	1974
SH 62	Drift, Jackson and Allen	1974–75
SH 63	Drift, Jackson	1974–75
SH 72	Drift, Allen and Jackson	1974–75
	Local solid remapping, Allen and Tappin	1973–75
SH 73	Drift, Jackson	1974–75
	Local solid remapping, Allen and Jackson	1972–74

GEOLOGICAL SUCCESSION

The geological formations shown on the 1:10 560, 1:25 000 and 1:50 000 maps are summarised below. The Tertiary and Mesozoic and the ?Precambrian rocks were encountered only in boreholes.

Quaternary

Blown sand
Peat
Lacustrine deposits
Alluvium
River terrace deposits
Alluvial fan
Marine and estuarine alluvium
Storm gravel beach deposits
Scree
Head
Fluvio-glacial gravel
Boulder clay
Fluvio-glacial terrace deposits
Morainic drift

Unconformity ∿∿∿∿∿∿∿∿∿∿∿∿∿∿∿∿∿∿∿∿∿∿∿∿

Tertiary (in Mochras Farm Borehole)

Oligocene	Interbedded silt and clay with sandy beds and thin lignites; conglomeratic in lower part

Unconformity ∿∿∿∿∿∿∿∿∿∿∿∿∿∿∿∿∿∿∿∿∿∿∿∿

Jurassic (in Mochras Farm Borehole)

Lias	Grey, calcareous mudstone and siltstone with limestone beds

Triassic (in Mochras Farm Borehole)

Dolomite and limestone with interbedded calcitic siltstone and sandstone

Unconformity ∿∿∿∿∿∿∿∿∿∿∿∿∿∿∿∿∿∿∿∿∿∿∿∿

Folding, regional metamorphism during Caledonian orogeny; extensive quartz-gold-sulphide veins after folding

Ordovician

INTRUSIVE ROCKS	Dolerite, rhyolite and quartz-microdiorite, mostly concordant or semi concordant intrusions contemporaneous with volcanism. Some dolerite dykes
ARAN VOLCANIC GROUP	
Aran Fawddwy Formation	acid ash-flow tuffs
Pistyllion Formation	mudflow breccias with intercalated lavas, tuffs and mudstone
Craig y Ffynnon Formation	acid ash-flow tuffs
Benglog Volcanic Formation	crystal tuff, intercalated pillow lavas, hyaloclastite and black mudstone. Thick silty mudstone and basalt flow locally above it
Melau Formation	bedded agglomeratic tuff and tuffite locally developed within silty mudstone and interbedded tuffite

Brithion Formation	acid ash-flow tuff locally developed at approximately same stratigraphical level as Melau Formation
Offrwm Formation	acid ash-flow tuff
Allt Lŵyd Formation	grey volcanic sandstone overlying interbedded black siltstone and sandstone with coarse quartzose sandstone (Garth Grit Member) at base. Aran Boulder Bed locally at top
Unconformity; local folding	∿∿∿∿∿∿∿∿∿∿

Cambrian

Intrusive rocks	Dolerite, microdiorite, quartz-microdiorite and microtonalite forming sills, dykes and laccoliths; intrusive breccias contemporaneous with Rhobell Volcanic Group. Porphyry-style copper mineralisation
Rhobell Volcanic Group	basalt with flow autobreccia and epiclastic breccias; laharic flows locally at base; rare fluviatile intercalations
Local unconformity, folding	∿∿∿∿∿∿∿∿∿∿
Mawddach Group	
Cwmhesgen Formation	dark grey and black mudstone, silty mudstone and siltstone; thin tuffaceous beds
Ffestiniog Flags Formation	thinly interbedded pale grey coarse quartzose siltstone and grey silty mudstone
Maentwrog Formation	grey silty mudstone with thinly interbedded coarse quartzose siltstone and fine sandstone and greywacke, mainly in the lower part
Clogau Formation	black silty mudstone
Harlech Grits Group	
Gamlan Formation	interbedded grey and purple siltstone, mudstone and thick beds of coarse-grained greywacke; some manganiferous horizons, and tuffitic beds in the upper part
Barmouth Formation	thickly bedded, coarse-grained greywacke with siltstone intercalations
Hafotty Formation	grey siltstone and silty mudstone with coarse-grained greywacke beds and manganese ore-bed near the base
Rhinog Formation	thickly bedded, coarse-grained greywacke with siltstone intercalations
Llanbedr Formation	grey, green and purple mudstone with siltstone and sandstone interbeds
Dolwen Formation	greenish grey siltstone and thinly bedded sandstone with conglomerate and pebbly sandstone at the base (seen only in Bryn-teg Borehole) and thin tuffaceous beds at top
Unconformity	∿∿∿∿∿∿∿∿∿∿

?Precambrian (in Bryn-teg Borehole)

Bryn-teg Volcanic Formation	interbedded sedimentary and volcaniclastic rocks, tuffite and andesitic lavas. Basalt veins are possibly contemporaneous

NOTES

All grid references refer to National Grid Square SH

Letters preceding specimen numbers refer to Geological Survey collections as follows:

E	Thin section, England and Wales Sliced Rock Collection
RU, GSM, Zs, HN, RX	Palaeontological specimens
HEXD	X-ray diffractometer trace, Leeds
NW	Specimen held in Isotope Geology Unit, London

Numbers preceded by the letter L refer to photographs in the official collection of the Geological Survey

Fossils held in the collections of the British Museum (Natural History) are numbered with the prefix BM(NH).

The authorship of fossil species is given in the Index of fossils.

CHAPTER 1

Introduction

GEOGRAPHY

The district (Figure 1) is mostly within the Snowdonia National Park. It is mountainous, partly forested and, inland of the coastal strip, sparsely populated. The main towns are Barmouth, Harlech, Trawsfynydd and Dolgellau.

The principal physical feature is the Rhinogs mountain range (Figure 2) with Y Llethr, at 754 m, the highest peak. In the east of the area the rounded Rhobell Fawr (734 m) is the highest point. The ground rises gradually from the morfas along the coast to the crest of the Rhinog range, but seen from the east the range is rugged and in sharp relief. Apart from the upper part of the Eden valley, which is broad and flat, the relief is generally high east of the range, and the deep steep-sided valleys of the Mawddach, Wen and Cwm-mynach contribute greatly to the scenic beauty.

Most of the district lies within the catchment of the Afon Mawddach and its major tributaries the rivers Eden, Gain, Gamlan, Wen and Wnion. The rivers Ysgethin, Cwmnant-col and Artro flow to the sea on the western side of the Rhinogs, and the Afon Prysor empties into Traeth Bach via Llyn Trawsfynydd, the largest lake in the district. Only the Afon Lliw in the north-eastern corner of the district drains eastwards, eventually to join the Dee.

Hill farming, forestry and tourism are the main industries but in the past quarrying and metal mining were prominent. The district contains the Dolgellau gold-belt, and mines are known to have been worked here at least since the early 18th century. About 140 000 oz gold were won mostly from vein deposits, together with an unknown quantity of copper, lead, zinc and silver up to 1916 after which production all but ceased. Manganese mining, however, continued until 1928.

PREVIOUS RESEARCH

Apart from some references to mining, systematic geological research in North Wales did not begin until the early 19th century and it was not until 1832, when Sedgwick first came to the area (Sedgwick and Murchison, 1835), that the rocks of western Merioneth were examined. Sedgwick published his findings on a map covering the whole of North Wales in 1845 and a year later Daniel Sharpe published a larger scale geological map of the same area. The Geological Survey began working in North Wales in 1846 and the geological map (Old Series one-inch Sheet 75 SE) of the area around Harlech was published in 1855. Some years later cross-sections across the Aran mountains (No. 29) and the Harlech Dome (No. 37) were published at six inches to one mile.

Because of the discovery of gold in 1846 geological investigations in the district have always had an economic as well as stratigraphical bias. Many publications are referred to in the text. Perhaps the most significant are by Sedgwick (1844), Belt (1867 and 1868), and Matley and Wilson (1946) on the stratigraphy; Readwin (1862) and Andrew (1910) on the gold-bearing veins; Wells (1925) on the volcanic and intrusive rocks; and Woodland (1939) on the manganese deposits.

GEOLOGICAL HISTORY

The oldest rocks proved in the district were in the Bryn-teg Borehole where sedimentary, volcaniclastic and calc-alkaline volcanic rocks underlie the Cambrian succession. They are

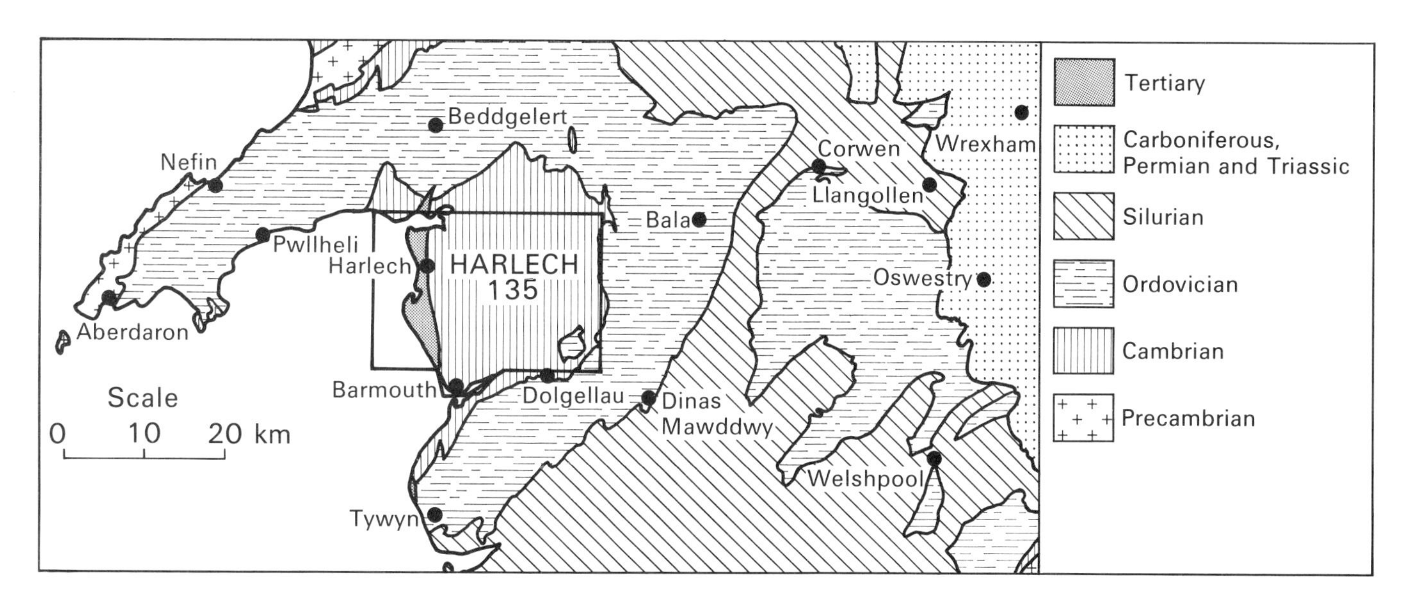

Figure 1 Location of Harlech (135) Sheet and simplified geology of part of North Wales

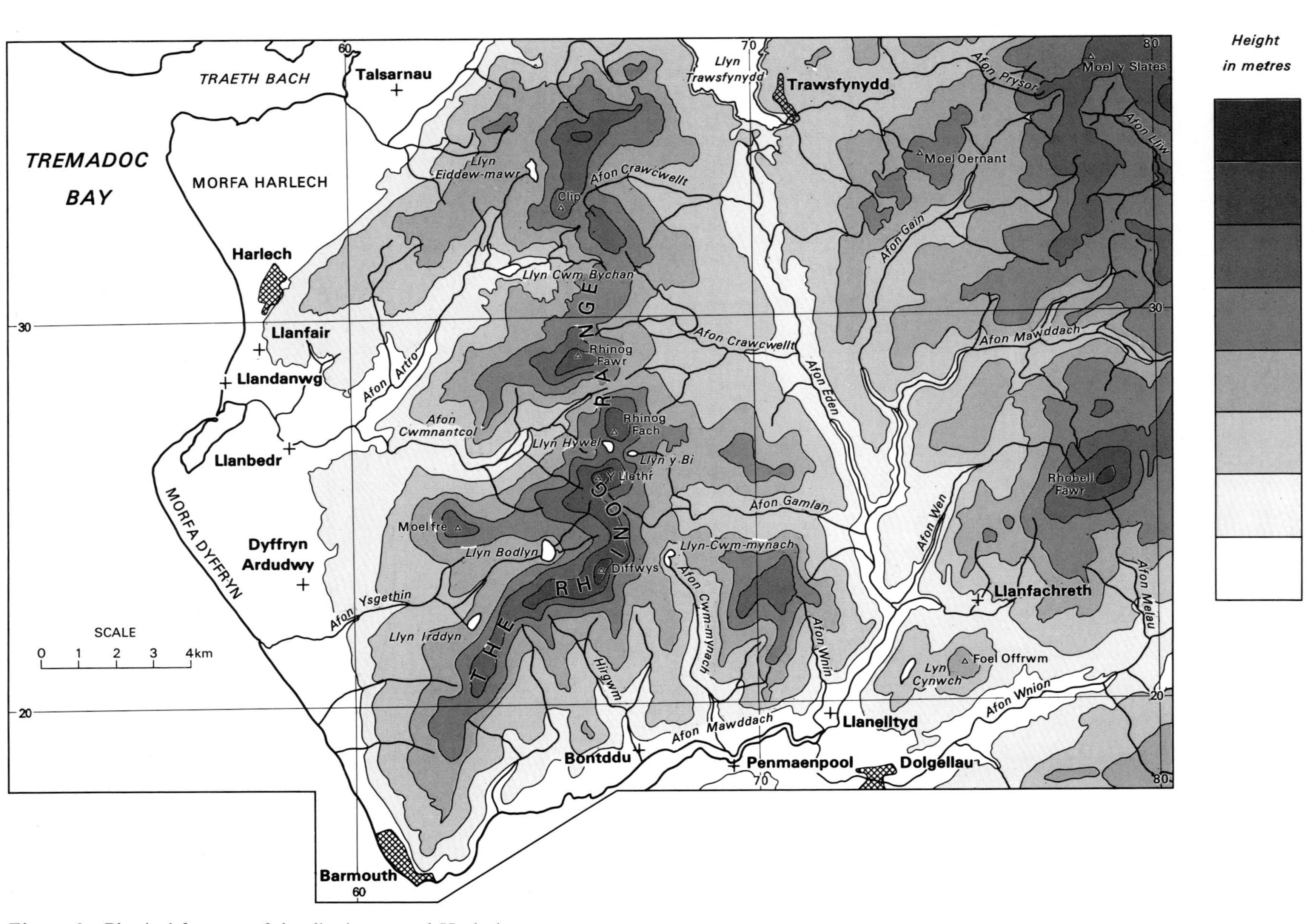

Figure 2 Physical features of the district around Harlech

believed to be of late Precambrian age and are in a typically island-arc succession. A comprehensive account has been given by Allen and Jackson, (1978), and is not repeated in the present work.

The Cambrian rocks are divided into the Harlech Grits, Mawddach and Rhobell Volcanic groups. The Harlech Grits and Mawddach groups, which together are about 4.5 km thick and are bounded by unconformities, contain no major depositional breaks and together span nearly the whole of the Cambrian period. All except the lowest 230 m, which were intersected in the Bryn-teg Borehole, are exposed. The two groups represent a complete cycle of basin formation and infilling within which there is an upward change from delta to prodelta, basin and open-shelf facies (Figure 3).

The Dolwen and Llanbedr formations represent the deltaic and prodelta facies in a generally fining-upward succession which contains, in a thin bed of tuff at the top of the Dolwen Formation, evidence of the oldest period of volcanism known in the Cambrian of the district.

The overlying basinal part of the succession is about 2.4 km thick. The lower part is characterised by coarse-grained, thickly bedded proximal turbidites interbedded with green, grey or purple silty mudstone. The thickest succession of turbidites is represented by the Rhinog Formation, some 425 to 780 m thick, which is interpreted as the channelised deposits of the mid to inner part of a submarine turbidite fan. Two main units of manganiferous rocks occur, one near the base of the Hafotty Formation, and the other in the upper part of the Gamlan Formation. In the latter, volcaniclastic rocks are associated with thin beds and laminae of manganiferous rocks which, it has been argued (Woodland, 1939), formed in a shallow, partly enclosed basin.

The Clogau Formation, which consists of black carbonaceous mudstone, marks a hiatus in turbidity current activity. A faunal break above it may indicate a period of non-deposition or emergence. When deposition from turbidity flows resumed the detritus carried into this part of the basin was uniformly much finer-grained than in the succession below. The turbidite beds are thin and have the characteristics of distal deposits. A long period of quiescent deposition, mainly of mudstone, led to the end of the basinal stage.

The open shelf deposition is marked by three distinct facies. The first, represented by the Ffestiniog Flags Formation, comprises thinly interbedded coarse quartzose siltstone and silty mudstone, laid down in a turbulent environment, possibly above wave-base. This was followed by quiescent deposition of black carbonaceous mudstone, the Dolgellau Member of the Cwmhesgen Formation, which in turn was followed by siltier mudstones laid down in better oxygenated conditions with more current activity. There are several breaks in deposition in the Cwmhesgen Formation and a record of intermittent volcanic activity throughout the upper part.

On the south-eastern side of the Harlech Dome the two upper formations of the Mawddach Group are unconformably overstepped by the Rhobell Volcanic Group. The pre-Rhobell uplift was accompanied by faulting and folding but north of the district, in the Migneint, Lynas (1973) observed a gradual passage from the Tremadoc to the Arenig. The Rhobell Volcanic Group is absent in the Migneint, and the severity of the unconformity in the south-east Harlech Dome can in part at least be attributed to tectonic disturbances associated with volcanism.

The Rhobell Volcanic Group is a remnant of a subaerially erupted volcano built almost entirely of basalt. The centre of eruption, which lies to the west of Rhobell Fawr, is marked by a complex of N-trending basic dykes and there are several high-level concordant dolerite intrusions around the outcrop of the lavas. At lower levels, however, the associated intrusions are dominantly acid to intermediate, and include a number of large laccoliths of microtonalite in addition to sills of microtonalite, quartz-microdiorite, microdiorite and dolerite. Dolerite in the dyke swarm across the central part of the Harlech Dome has strong geochemical affinities with the rocks in the Rhobell Volcanic Group. The mainly NW and subsidiary NE-trends of the dykes attest to a period of uplift creating tensional forces at this time. Several pipes and many dykes and sills of intrusive breccia are attributable to this magmatic event which is also responsible for the disseminated copper deposit at Capel Hermon and the copper deposit worked in the Glasdir breccia pipe.

The construction and partial destruction of the Rhobell Fawr volcano took place within a short time interval near the end of the Tremadoc epoch. The volcanic rocks were then gently folded, and both they and the surrounding areas of exposed sedimentary rocks of the Mawddach Group were submerged by the transgressive Arenig sea.

The Aran Volcanic Group, which overlies unconformably the Mawddach and Rhobell Volcanic groups, records a volcanic episode that lasted from Arenig to Caradoc times. The group comprises a bimodal suite of magmatic rocks and interbedded sedimentary rocks.

The basal Allt Lŵyd Formation consists of shallow-water, mostly clastic, sedimentary rocks. The composition of the coarse quartzose sandstone in the Garth Grit Member at the base suggests derivation from uplifted Harlech Grits Group to the west, but the provenance of the sandstones above it is almost entirely volcanic. Small islands of Rhobell Volcanic Group probably remained in the Arenig sea, but most of the detritus in the Allt Lŵyd Formation appears to have been derived from contemporaneous volcanism.

Above the Allt Lŵyd Formation are seven further formational divisions of the Aran Volcanic Group, tending to reflect a general alternation of basic and acid eruptive episodes. The volcanic rocks are intercalated throughout with black silty mudstone, which in places is tuffitic and contains rare thin beds of oolitic ironstone. The group is thickest in the south and the reduction in thickness northwards is accounted for in part by wedging out of the lower formations. The basic rocks, which include spilitic pillow lavas, hyaloclastites, and associated tuffs and tuffites, tend to represent local submarine eruptions, whereas the acid ash-flow tuffs form thick units some of which persist over great distances and probably emanated from one emerged eruptive centre to the south-east of the district. Contemporaneous intrusions are mostly of dolerite, forming concordant and semi-concordant sheets within the group and in the upper parts of the Cambrian sub-volcanic basement. Some dolerite dykes in the Cambrian show the same geochemical characteristics as these sills.

Four phases of folding can be identified marking progressive stages of deformation during the Caledonian orogeny. Two minor, roughly co-axial (N to NNE-trending), local phases preceded the Arenig transgression. There is evidence that the initiation of the main structures in the Harlech Dome dated from this time. A third phase, trending NNW, is possibly related to earth movements at the end of the Ordovician. The main folds, however, are attributable to the climactic end-Silurian deformation. They are broad, open, N or NE-trending periclinal structures with a local axial-plane cleavage. Accompanying regional metamorphism nowhere exceeds lower greenschist facies. Faults that were active at least as early as late Cambrian can be identified, and there is evidence of fault control on sedimentation during the Ordovician. The Bala Fault, which is interpreted as a major wrench structure, crosses the south-east corner of the district and was active a number of times during the Lower Palaeozoic.

A number of quartz-sulphide-gold veins, which were worked profitably for gold in the 19th century, were emplaced immediately after the climax of the Caledonian folding.

The Mesozoic and Tertiary rocks are nowhere exposed in the district, but were encountered in the Mochras Farm Borehole beneath a cover of Quaternary sediments. A thick succession of late Triassic, early Jurassic and Oligocene sediments, described in Woodland (1971), was encountered and not bottomed to the west of the Mochras Fault in the coastal area. The Triassic sediments lie on the margin of the Cardigan Bay Mesozoic basin. It is not known whether they or the marine early Jurassic sediments encroached eastwards over the Harlech Dome or whether the bounding Mochras Fault was continuously active throughout deposition, though the predominantly fine-grained nature of the Jurassic sediments in the borehole suggests that the line of the fault was not the contemporary basin margin. The early Oligocene deposits contain thick conglomerates and with the later beds are considered to be those of a floodplain parallel to a subdued fault-scarp.

Quaternary deposits resulting from the most recent glaciation are widespread and varied. They suggest that the entire district was covered by ice at the height of at least one of the Pleistocene glaciations, and deposits of locally derived till or boulder clay are widespread.

CHAPTER 2

Cambrian rocks

Rocks of Cambrian age (including those of the Tremadoc epoch) comprise the greater part of the district. They were first investigated by Sedgwick and attributed to the Cambrian System by Sedgwick and Murchison when it was defined in 1835. Subsequent work by the Geological Survey and other geologists led to progressively more detailed subdivision of the Cambrian succession. Historical reviews, referring to the most important stages in the evolution of the classification and nomenclature of the Cambrian rocks, are given by Andrew (1910), Matley and Wilson (1946), Rushton (1974), and Allen, Jackson and Rushton (1981).

The Cambrian succession (Figure 3) consists of clastic sedimentary and volcanic rocks, divisible into three major groups. In ascending order they are the Harlech Grits Group, a formal term for the long established Harlech Grits; the Mawddach Group, a new term defined by Allen, Jackson and Rushton (1981) to include all the sedimentary strata above the Harlech Grits Group; and the Rhobell Volcanic Group redefined by Kokelaar (1977, 1979) after Wells (1925) to include the extrusive rocks which lie between the Mawddach and Aran Volcanic groups. The formational divisions of the groups are shown in Table 1.

The base of the Cambrian succession is not easily determined. In the Bryn-teg Borehole, Allen and Jackson (1978) described about 210 m of sedimentary rocks attributable to the Dolwen Formation below the stratigraphically lowest exposure. These rocks rest with probable unconformity on about 160 m of volcanic rocks, which they designated the Bryn-teg Volcanic Formation. There is no evidence of age of any rocks below 71.92 m in the borehole, where Rushton (*in* Allen and Jackson, 1978) found *Platysolenites antiquissimus*, a Lower Cambrian foraminifer. The Bryn-teg Volcanic Formation contains andesite and dacite lavas, coarse tuffitic mass-flow deposits, graded siltstone, fine volcanic sandstone,

Table 1 Chronostratigraphical and lithostratigraphical subdivisions of the Cambrian

Series	Group	Formation	Member
Tremadoc	Rhobell Volcanic		
Merioneth	Mawddach	Cwmhesgen	Dol-cyn-afon Dolgellau
		Ffestiniog Flags Maentwrog Clogau	
St David's			
	Harlech Grits	Gamlan Barmouth	
Comley		Hafotty Rhinog Llanbedr Dolwen	

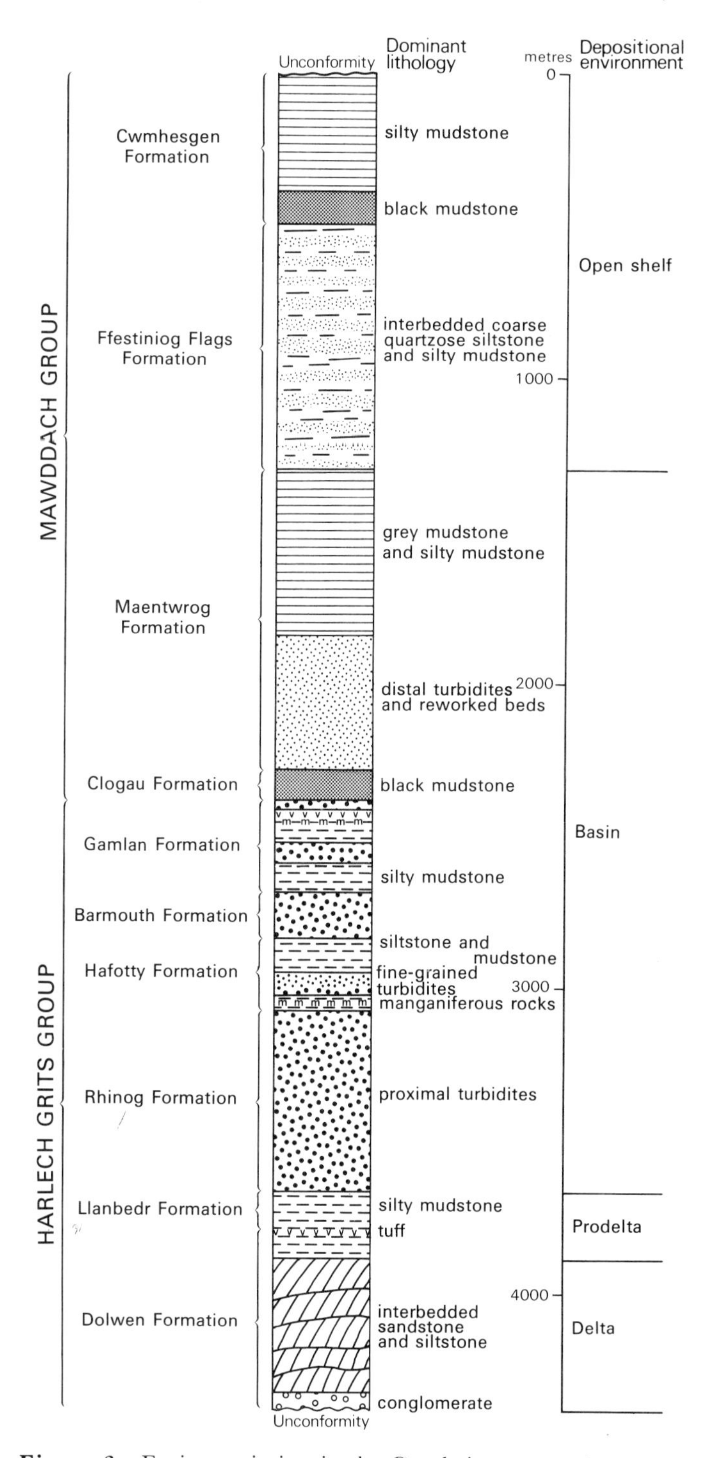

Figure 3 Facies variation in the Cambrian succession

thin mudstone, and rare laminae of air-fall tuff, intruded by veins of basalt. The formation, which is essentially comparable to a submarine island-arc sequence, is unlike the Arvonian, which Allen and Jackson (1978) argue is probably younger than the Bryn-teg Volcanic Formation. It is generally assumed that the Arvonian is late Precambrian, but an age in the Lower Cambrian cannot yet be ruled out either for these rocks or the Bryn-teg Volcanic Formation.

In general terms the main characteristics of the three groups above the Bryn-teg Volcanic Formation have long been known. The Harlech Grits Group, which consists mainly of coarse-grained greywacke interbedded with siltstone, was recognised as a distinct geological unit by Sedgwick and Murchison (1835). It was shown with some precision as 'Cambrian sandstone, slate exceptional' on the Geological Survey map (one-inch sheet 75 SE) of 1854. On the same map the 'Lingula Beds' is equivalent to the Mawddach Group, and consists mainly of thinly interbedded dark grey silty mudstone and pale grey fine sandstone or coarse quartzose siltstone. The term 'Lingula Flags' was later used in a more restricted sense to represent the part of the succession above the Clogau Formation and below the Tremadoc Slates. The Rhobell Volcanic Group, however, was not firmly established as extrusive volcanic in origin until late in the 19th century (Ramsay, 1881), and its Cambrian age was not demonstrated until the work of Cole and Holland was published in 1890.

HARLECH GRITS GROUP

It is proposed here to define formally as a group the sequence informally called the Harlech Grits. In the classification scheme of Matley and Wilson (1946) the group includes the Dolwen Grits, Llanbedr Slates, Rhinog Grits, Manganese Group, Barmouth Grits, Gamlan Flags and Grits, and Cefn Coch Grit. Each of these units, except the Cefn Coch Grit, is recognised as a formation and defined. The Cefn Coch Grit has been included, unnamed, within the Gamlan Formation. An examination of the base of the overlying Clogau Formation shows that there is no easily and consistently recognisable bed, or group of beds, of sandstone to justify the retention of a named unit at this stratigraphic level.

The group is extensively exposed in the central area of the Harlech Dome (Figure 4) and in two small inliers—the Llafar fold inlier and Bryn Celynog faulted inlier—to the north-east of the dome. Only small parts of these areas have been remapped by the authors. The linework on the map and the essential definitions of the formations, therefore, are those of Matley and Wilson (1946).

Dolwen Formation (575 m)

Approximately the upper half of this, the lowest formation in the group, is exposed in the core of the Dolwen pericline. The lower half was intersected in the Bryn-teg Borehole [6992 3214]. In their description of the borehole and surrounding area Allen and Jackson (1978) recognised ten lithological subdivisions of the formation (Table 2).

At the base are conglomerates with sparse, intercalated beds of pebbly sandstone. Above them the pebbly sandstones form a transitional unit into a sequence of interbedded sandstone and siltstone which displays sedimentary characteristics (Plate 1.1) comparable to modern sediments from fluviatile, delta or open shelf environments. Several phases of delta building are recognisable. The topmost unit marks the beginning of a change in sedimentation style indicative of deepening of the basin and a shift from delta to more distal, probably prodelta parts. In this unit silty mudstone, uncommon below, is the dominant lithology, and there are thin graded beds of greywacke which are probably turbiditic in origin.

A number of important changes in clastic content are evident at different levels in the formation. In the basal conglomerates the pebbles and cobbles are mostly of volcanic rocks compositionally different from those in the underlying Bryn-teg Volcanic Formation. The overlying pebbly sandstones contain, in addition to volcanic rocks, locally abundant sand-size granitic fragments and in places up to 60 per cent quartz in the matrix. Detrital magnetite, which at higher levels in the formation forms thin laminae, is also locally abundant. The dominantly lithic character of the clastic component persists until the subgreywackes and protoquartzites (Pettijohn, 1957) are replaced by feldspathic greywacke (locally arkose) and subarkose (Table 2).

At the top of the formation, Matley and Wilson (1946) commented on the presence of distinctive purple or lavender sandstones. Forming a unit 1.5 to 2 m thick, they are exposed in a number of localities including both rivers

Table 2 Lithological subdivisions of the Dolwen Formation

Subdivision	Main rock types	Thickness in metres	
silty mudstone, sandstone at top	Feldspathic greywacke and subarkose	80	Exposed
sandstone with siltstone bands		55	
medium-grained sandstone with some coarse beds		65	
fine-grained sandstone with some medium-grained beds		60	
siltstone	subgreywacke and proto-quartzite some lithic greywacke in coarse siltstone	65	
sandstones		75	Borehole only
siltstone		60	
sandstones		70	
pebbly sandstones	Volcanic conglomerates and lithic greywacke	20	
conglomerates		25	

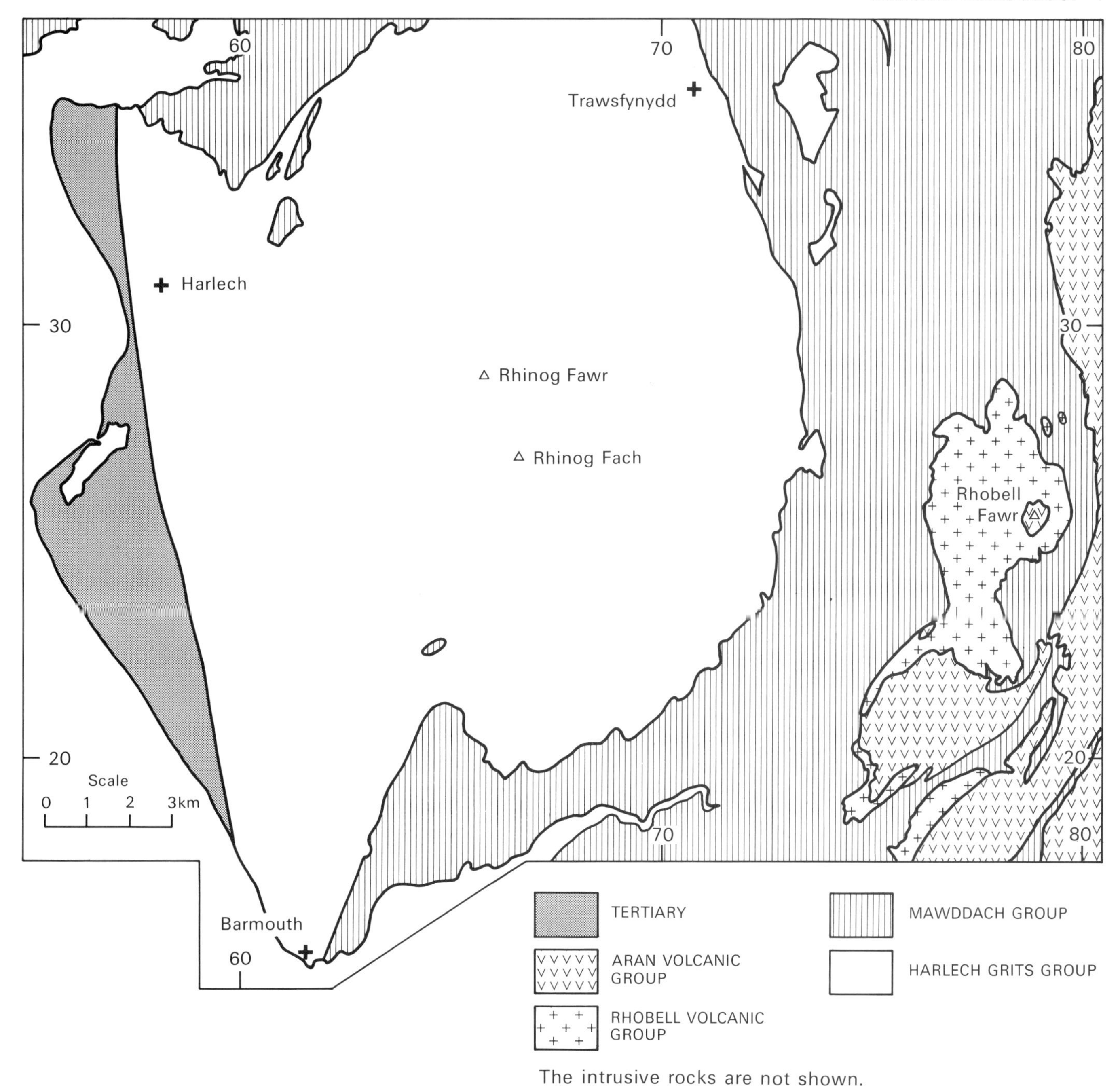

Figure 4 The major geological divisions on the Harlech (135) Sheet

Crawcwellt [7082 2889; 6903 3303]. At the southernmost of these two localities interbedded siltstone contains laminae of euhedral feldspar crystals, fragments of quartz euhedra and some chlorite/muscovite pseudomorphs (E 50282) clearly of volcanic origin, and possibly air-fall tuff. This is the first record of contemporaneous volcanism in the lower Cambrian of this area. Laminae in associated sandstone beds are convoluted.

Llanbedr Formation (90 – 180 m)

This formation, which contains the best workable slates in the district, is exposed around the Dolwen Formation in the central area and in a largely drift-covered strip near Llanbedr. Matley and Wilson (1946) estimated the thickness to be 90 m in the central area and 180 m at Llanbedr. The main rock types are purple, bluish grey or green mudstone, commonly with a good slaty cleavage, and broadly spaced beds of green siltstone, in a succession that compares with that commonly found in a prodelta environment. Beds of sandstone occur sparsely in the lowest parts of the formation in the central area. Near Llandbedr, sandstone is regularly interbedded with cleaved sandy mudstone above the workable slates. This accounts for the greater thickness of the formation. Bates (1975) described sandstone dykes in these rocks.

Plate 1

1 Cross-bedded unit in sandstones in the Dolwen Formation (L 1263).

2 Convolute bedding in a sandy bed in the Gamlan Formation (L 1266).

3 Folding in silty mudstone and sandstone of the Gamlan Formation on the foreshore, Barmouth (L 1264).

1

2

3

The conformable and transitional nature of the junction with the underlying Dolwen Formation is demonstrable in the southern Afon Crawcwellt [7082 2889] about 90 m upstream from Pont-Llyn-y-cafn. Here, the topmost 2 m of the Dolwen Formation consist of thinly interbedded green and purple laminated sandstone and purplish grey siltstone with volcaniclastic laminae. Immediately overlying these rocks is purple mudstone taken here to define the base of the Llanbedr Formation, and in the lowest 35 m of the formation exposed here there are rare beds, up to 30 cm thick, of green coarse sandstone, and a 4 m-thick unit of green siltstone interbedded with the mudstone. The base of the formation may also be examined in the northern Afon Crawcwellt [6903 3303], where the purplish grey mudstone shows numerous green reduction spots.

Rhinog Formation (425–780 m)

The formation consists mainly of grey, greyish green or bluish grey greywacke with thin intercalations of siltstone and mudstone. There are thin beds, lenses or multiple units of coarse sandstone or quartz pebble conglomerate, some of which, according to Matley and Wilson (1946), can be traced for up to 1.5 km. Throughout the succession there are beds of well sorted, coarse-grained, quartzose sandstone.

The base of the formation is defined in Ffridd-Lŵyd on the south of Trawsfynydd Lake [6859 3417] at a point where greenish grey cleaved mudstone of the Llanbedr Formation is conformably overlain by a bed of pale green greywacke, the lowest of a thick succession of coarse-grained turbidites exposed in the crags.

The formation is thickest in a belt trending north-east from Barmouth, and thins to the south-east and north-west, though there is some confusion about the position of the base of the formation and, therefore, its thickness in the latter direction. Matley and Wilson (1946) themselves were unsure about the inclusion in the Llanbedr Formation rather than the Rhinog Formation of the sandstone and siltstone sequence above the workable slates in the coastal area and the beds clearly mark the advance of the prograding turbidite fan.

The greywacke is medium- to coarse-grained, and forms beds up to 4 m thick. Woodland (*in* Matley and Wilson, 1946) gave a petrographic description of the rocks which indicates a range of composition from feldspathic greywacke to subarkose using the nomenclature of Pettijohn (1957). Coarse-grained rocks commonly contain a prevalence of granitic fragments over feldspar, putting them in the fields of lithic greywacke or protoquartzite. In all rocks, however, clasts of quartz, commonly blue or pink, are dominant, and among the greywackes the term quartz wacke (Krumbein and Sloss, 1963) would be appropriate.

Kuenen (1953) and Kopstein (1954) were the first to interpret these rocks as turbidites. Most beds are graded, with a few showing a basal layer of pebbles, and have a parallel laminated upper layer. It is not uncommon for the whole of a bed to show parallel lamination or for amalgamated units to occur. Other combinations of the elements defined by Bouma (*in* Dzulynski and Walton, 1965), notably *abc* or *ae* are less common. In addition to washouts and burrows, a wide variety of sole marks has been observed in these rocks by Crimes (1970) including flute and groove casts and load structures. Inclusions of mudstone, siltstone and, less commonly, laminated sandstone have also been recorded. Crimes (1970) concluded that the greywackes are proximal turbidites and that sole markings indicate derivation from the north.

Beds from 5 to 30 cm thick of coarse-grained, well-sorted sandstone, many with quartz pebbles, were believed by Crimes (1970) to have been derived by reworking of clastic material laid down by turbidity currents. Some display cross bedding. Washouts at their bases show a preferred east–west orientation and are commonly filled with pebbles, in places, predominantly of cleaved acid volcanic rocks. Interbedded siltstone is a minor component of the formation. In places interbeds display complex slump structures which are truncated by the overlying turbidite.

The formation is the thickest succession of turbidites in the Harlech Grits Group and, displays the characteristics of classical turbidites corresponding mainly to Facies C of Ricci-Lucchi (1975). They are proximal turbidites and may be interpreted as mainly channelised deposits of the mid to inner fan.

Hafotty Formation (170–300 m)

This is the Manganese Group of Matley and Wilson (1946). The formation was originally named Hafotty Group (*in* Andrew, 1910) after the principal manganese mine, and later renamed Hafotty Formation by Cowie, Rushton and Stubblefield (1972). Woodland (1938, 1939) gave the first detailed description of the formation and reviewed earlier references to it. Geochemical studies are reported by Mohr (1956, 1959, 1964), Mohr and Allen (1965) and Glasby (1974). Kopstein (1954) examined the sedimentary structures in the sandstone beds.

Matley and Wilson (1946) divided the formation into three members:

Upper Manganese Shales	100–200 m
Manganese Grit	2– 60 m
Lower or Ore-bed Shales	15– 20 m

However, the ore-bed has not been recognised in the outcrop north of Y Garn [703 230]. The Manganese Grit thickens eastwards across the dome and the interbedded sandstones in the Upper Manganese Shales become more numerous and thicker north-eastwards from Barmouth. North of Y Garn there is no certainty about the definition of the base of the formation and it may have been placed above the equivalents of the Manganese Grit.

The Hafotty Formation follows conformably the Rhinog Formation. A basal contact, here taken to define the formation, is exposed at the Hafotty Mines [6164 1830] on the hillside above Llanaber. The coarse-grained, thickly bedded turbidites of the Rhinog Formation pass abruptly into grey, thinly bedded, fine-grained sandstone and siltstone of the Lower or Ore-bed Shales. These were divided by Woodland (1939) into six sub-units:

	m
Mudstone	10.0–15.0
'Bluestone Grit'	0.6– 1.2
'Bluestone'	1.8
Manganese ore-bed	0.3 (average)
Pyritous mudstone	0.05
Mudstone	10.0

The lowest mudstone unit is uniform in thickness over all the area in which the ore-bed is recognised and is not noticeably manganiferous though according to Mohr (1959) it contains slightly more MnO than the Llanbedr Slates. The ore-bed, which always overlies a bed of pyritous mudstone, is a very fine-grained, compact rock showing chocolate-red, yellow and, less commonly, bluish black bands when fresh. The red colour is attributable to hematite dust. At surface the ore weathers to manganese oxide, in places botryoidal, identified by Glasby (1974) as todorokite. The constituent minerals of the ore-bed are rhodochrosite (dialogite), spessartine, quartz, minor magnetite, hematite, chlorite, pyrite and rare rhodonite.

Above the ore-bed the formation consists mostly of banded siltstone, mudstone and fine sandstone showing cross and parallel lamination and could be interpreted as distal turbidites. Medium to coarse-grained greywacke in graded beds up to 60 cm thick with laminated and cross-laminated silty tops commonly forms the hanging wall of the ore-bed workings. Above this the Manganese Grit is the thickest and most persistent unit of greywacke, but a few beds occur throughout the Upper Manganese Shales, particularly near the top.

Woodland (1939) considered that sedimentary structures within the ore-bed, including banding, uniformity of grain size, spherulitic texture, distortion of banding and contraction cracks, indicate an origin by precipitation as a colloidal gel, probably of rhodochrosite with clay and silica, simultaneously with iron hydroxides. The spessartine is a result of regional metamorphism. Both he and Mohr (1956), who used trace element geochemistry to rule out a volcanic source, favoured a source for the manganese in a gneissic landmass. Geochemical studies of the constituent minerals carried out by Glasby (1974) support a diagenetic origin for the ore-bed and rule out the need for a manganese-rich provenance. He suggested that the ore formed in a shallow marine basin in which reducing conditions had developed. Under such conditions manganese in the sediment column is extensively remobilised and deposited at the sediment-water interface as rhodochrosite. Fractionation during this process leads to the precipitation of iron as sulphide in the underlying sediment.

The configuration of such a shallow enclosed basin is difficult to define, because the 'Bluestone Grit' and the sandstones in the Manganese Grit and above it are turbiditic greywackes similar to those in the underlying Rhinog and overlying Barmouth formations, but the rapid accumulation of the Rhinog Formation in a submarine valley may have raised rapidly and temporarily the local level of the sea bed.

Barmouth Formation (60 – 230 m)

Consisting of greywackes showing the characteristics of proximal turbidites, this is similar to the Rhinog Formation and likewise shows a maximum thickness in an elongate NNE-trending central part of the area, thinning towards the north-west and east. Matley and Wilson (1946) observed that there were fewer mudstone intercalations than in the Rhinog Formation, and both they and Crimes (1970) commented that it was generally coarser.

The Barmouth Formation follows conformably the Hafotty Formation. The proposed stratotype base is exposed by a footpath [6160 1580] which runs along the hillside above Barmouth. It is marked by the abrupt appearance of coarse, pebbly greywackes. The lowest bed shows the ideal sequence described by Bouma (intervals *abcd* and *e*). Many of the other characteristic features of turbidites, including thick beds (up to 2 m), simple, reverse and multiple grading, and load and flame structures, are displayed in this section. Higher in the formation middle-absent and interrupted sequences are predominant. The sandstone to shale ratio is always high but a thin unit of banded siltstone is intercalated [6166 1584] in the lower part.

About four cycles of turbidite sedimentation are apparent in the Barmouth Formation. The beds become finer grained and thinner with proportionally more shale towards the top of each 'cycle'. Crimes (1970) described various sole marks at the bases of the turbidites and observed their north-north-west alignment.

Gamlan Formation (230 – 360 m)

The formation is exposed in the Caerdeon and Goedog synclines and on the eastern side of the Dolwen pericline; it is also present in folded and faulted inliers in the north-eastern parts of the district. It consists of greenish grey and colour-banded green, bluish grey and purplish grey silty mudstone, locally pyritic, and thin beds of coarse-grained sandstone (Plate 1.3) which occur at intervals throughout the formation in most areas. The conformable base of the formation is exposed in the Afon Gamlan, but access is difficult through afforested ground and the rocks are cut by intrusions. A more easily accessible site to define the base is in the steeply dipping beds above Barmouth. Just to the north of a footpath [6178 1584] the coarse beds of the Barmouth Formation are abruptly overlain by greenish grey silty mudstone, with thinly interbedded massive, cross- and parallel-laminated units. Beds showing convolute lamination (Plate 1.2) are less common. Beds of graded greywacke, 1 to 9 cm thick, and a bed over 1 m thick of coarse-grained greywacke can be traced laterally for some distance. One of these beds, showing an uneven, channelled base, crosses the path [about 6182 1584].

Within the predominantly silty mudstone succession Matley and Wilson (1946) identified four major sandstone members, including the Cefn Coch Grit. The beds are turbidite sandstone, greywacke to quartz wacke (Plate 3.2) in composition, up to 2 m thick, in places with pebbly bases; they resemble the sandstones in the lower formations.

In all parts of the area the upper half of the formation is manganiferous. The mine listed by Dewey and Dines (1923) at Ffridd-llwyn-Gurfal [6114 3088] lies within this formation, and Goodchild in an unpublished report dated 1893, quoted by Dewey and Dines (1923), regarded this formation as the Upper Manganese Bed. Spessartine (Plate 3.4), commonly with quartz and chlorite, forms beds, laminae, nodules, lenses, and structures that Matley and Wilson (1946) believed were infilled worm tubes, in most places interbedded with purplish grey mudstones. No beds thicker than 5 cm have been found. Associated beds of mudstone and coarse siltstone are usually rich in disseminated garnet. Garnetiferous laminae not uncommonly show evidence of contemporaneous, wet-sediment deformation. Price (1963) believed that the so-called worm tubes, in which he found rare pumpellyite, were inorganic in origin, and suggested that the spessartine rock had the same origin as the manganese ore-bed in the Hafotty Formation.

In the Afon Gain [7362 3152] and just north of the district boundary near Gellilydan [6883 3961] thin beds of tuffitic mudstone (Plate 3.3) occur near the top of the formation. At the Afon Gain locality fragments of embayed quartz crystals, brownish green chlorite forming irregular and rectiliniar shapes, some attached to quartz, and blade-like chlorite/magnetite fragments, all scattered through mudstone, are possibly of volcanic origin (E 48305). At this same stratigraphical level elsewhere, beds of microcrystalline quartz, sericite and chlorite (E 45907), though structureless, may be ashy in origin. Several thin, probably tuffaceous, beds are also present in the Barmouth area [6195 1595].

Coarse-grained turbidite sandstones persist to the top of the formation, but interbedded with them in the upper half are laminae and thin beds of the coarse quartzose siltstone and fine sandstone which characterise the Mawddach Group. They are rare except in the top 20 m where, in some areas, they exceed coarse sandstone in proportion. Near Clogau gold mine, [6753 2026], for example, there are beds of laminated quartzose sandstone 3 cm thick and fine-grained massive quartz wacke (E 54034) up to 20 cm thick at the top of the formation. PMA, AAJ

Biostratigraphy

Very few fossils have been found so there is much uncertainty about the precise correlation of the formations in the group. Fossils of stratigraphical value are known only near the base and top.

The only body fossil from the Dolwen Formation is a specimen of *Platysolenites antiquissimus* which was described from the Bryn-teg Borehole by Rushton *in* Allen and Jackson (1978, pp. 46–48) and interpreted as representing a low horizon in the Lower Cambrian. The single specimen is insufficient to allow definite correlation with the *P. antiquissimus* Zone of Estonia and hence with the upper Tommotian Stage of the Siberian succession. The occurrence suggests a marine environment because *Platysolenites* is elsewhere associated with typically marine fossils.

Inarticulate brachiopods have been collected from the Llanbedr Formation near Llyn Cwm Mynach (Lockley and Wilcox, 1979) but they are too poorly preserved to be of biostratigraphical value. Apart from a few burrows, no definite fossils have been collected from the Rhinog, Hafotty and Barmouth formations. However, if Nicholas' (1915) lithological correlation between the successions of the Harlech Dome and St Tudwal's Peninsula is correct, the top of the Rhinog Formation is probably of late Lower Cambrian (late Comley Series) age because a fauna of that age has been described from the top of the Hell's Mouth Grits (Bassett, Owens and Rushton, 1976). The Barmouth Formation has yielded a few burrows suggestive of bathyal waters (Crimes, 1970, p. 131).

Structures described as burrows, but believed by Price (1963) to be inorganic in origin, are so abundant in some beds of the Gamlan Formation that they were formerly spoken of as 'fucoid beds'. There are a few body fossils known from the Gamlan Formation: a fragment of *Paradoxides* (found loose) and a *Parasolenopleura*? (registered numbers RU 9705–6) from 0.5 km S of Gwynfynydd Farm suggest a middle St David's (Middle Cambrian) horizon. The transitional succession from Gamlan to Clogau Formation in the Afon Llafar yielded fragments of *Paradoxides hicksii* and *Eodiscus punctatus* s.l. (Allen, Jackson and Rushton, 1981, p. 303), and show that the top of the Gamlan Formation is referable to the *Tomagnostus fissus* Zone of the St David's Series. This record supports the correlation of the Gamlan–Clogau boundary with the boundary between the Upper Caered Mudstones and the Nant-pig Mudstones at St Tudwal's Peninsula made on lithological evidence by Nicholas (1915). AWAR

MAWDDACH GROUP

The Mawddach Group and its component formations (Table 1) were defined by Allen, Jackson and Rushton (1981). The group follows the Harlech Grits Group conformably and is unconformably overlain by both the Rhobell and Aran volcanic groups. It contains the Clogau Shales, Vigra Flags, Penrhos Shales, and Ffestiniog Flags of Matley and Wilson (1946), and the Dolgellau Beds and Tremadoc Slates of other authors; all are now reclassified into four formations. These formations, the Clogau, Maentwrog, Ffestiniog Flags and Cwmhesgen, are described in full by Allen, Jackson and Rushton (1981). Only brief accounts are given here.

Clogau Formation (90–105 m)

Equivalent to the Clogau Shales of Matley and Wilson (1946) the formation consists of dark grey or black banded carbonaceous mudstone (Plate 3.5) and silty mudstone with rare quartzose silty laminae and even less common beds of fine sandstone. Lenses and laminae of sulphides are ubiquitous.

The junction between the Clogau and Gamlan formations is transitional, and at the type locality in Afon Llafar [7357 3644] a transitional unit 5 m thick containing the interbedded diagnostic rock types of both the Harlech Grits and Mawddach groups, immediately underlies the Clogau Formation.

Maentwrog Formation (700–1200 m)

The quiescent depositional conditions suggested by the sediments of the Clogau Formation ended abruptly with the re-invasion of the basin by turbidites (Plate 2.1). They form beds 1 to 80 cm thick ranging from coarse quartzose siltstone to fine sandstone (Folk, 1974). In composition they are mostly quartz wacke with some greywackes (Plate 3.1) near the base. The turbidites are interbedded with grey silty mudstone and usually discontinuous laminae and thin beds of cross-bedded clean coarse-grained quartzose siltstone, presumably reworked turbidite material redeposited by persistent sea bed currents. As noted by Hsu (1964) such beds are common within turbidite successions.

The base of the Maentwrog Formation is defined near

Bontddu just upstream from Vigra Bridge [6682 1922], where the sudden appearance of turbidites on top of the Clogau Formation can clearly be seen. In most parts of the area turbidites occur only in the lower half of the formation, the upper part being composed almost entirely of mudstone. This led Matley and Wilson (1946) to divide the Maentwrog Flags into a lower Vigra and an upper Penrhos member. The distinction, however, is not tenable in all parts of the area. East of Trawsfynydd, for example, there are two thick arenaceous units within the formation. In the area south of Mynydd Mawr on the eastern side of the district, the simple passage of a lower dominantly arenaceous component into an argillaceous member is obscured by the presence of pale grey siltstone in the lower member.

Ffestiniog Flags Formation (650 – 1020 m)

This formation, which contains the Ffestiniog Group of Belt (1867) and the lower part of the *Parabolina* Beds of Fearnsides (1905), consists of massive or poorly bedded pale to dark grey silty mudstone and thinly bedded light grey coarse-grained quartzose siltstone ('ringers' of Fearnsides, 1910) (Plate 2.2). Each rock type is dominant in alternating units, up to 30 m thick, which grade into each other through transitional units in which the rock types are thinly interbedded in roughly equal proportion. Sedimentary structures throughout the formation are indicative of deposition above wave base in a shallow tidal or alluvial/estuarine environment.

The type section is defined in a tributary of the upper Afon Mawddach [7492 2947] near Bryn-y-gath. Here the base of the formation is taken at the first appearance of light grey silty mudstone in beds 5 to 8 cm thick within dark grey, thinly bedded mudstone with laminae of quartzose siltstone characteristic of the underlying Maentwrog Formation. The junction between these two formations is everywhere transitional.

Cwmhesgen Formation

The thick succession of mainly black and dark grey mudstone, silty mudstone and siltstone, which conformably follows the Ffestiniog Flags Formation, has been divided into as many as eight separate sub-units (Fearnsides, 1905) largely on faunal content. The twofold division into Dolgellau Beds and the overlying Tremadoc Slates (Belt, 1867) is the most general and can be upheld on lithological as well as faunal grounds. In defining the new Cwmhesgen Formation, therefore, Allen, Jackson and Rushton (1981) divided it into Dolgellau and Dol-cyn-afon members, which are more or less equivalent to the Dolgellau Beds and Tremadoc Slates.

The **Dolgellau Member** (63 to 150 m) consists of dark grey to black mudstone and silty mudstone. The black rocks are rich in carbonaceous material, and contain laminae and lenses of sulphides. Quartzose laminae and thin beds are present locally near the base only. Thin beds of fragmental rock (Plate 3.6) and finely laminated beds, both containing small black phosphatic nodules, possibly mark standstills in clastic deposition. There are thin beds of reworked tuff near the top of the member. The type section is exposed in the upper Afon Mawddach and the base is defined near its junction with Afon Cwmhesgen [7866 2935] where the Ffestiniog Flags Formation passes abruptly into flaggy, dark grey mudstone of the Dolgellau Member with rare thin graded beds and laminae of coarse quartzose siltstone. Elsewhere, the junction is less easy to define, and near Derwas [7037 1785] it is possible that the upper part of the Ffestiniog Flags Formation is laterally equivalent to the Dolgellau Member in other parts.

The **Dol-cyn-afon Member** (160 to 340 m) has a transitional junction with the Dolgellau Member. The base of the stratotype is defined in a tributary of the Afon Mawddach [7941 2873] at the bottom of a 47 cm-thick bed of banded dark grey mudstone overlying the dominantly black lithologies of the Dolgellau Member. The lowest occurrence of *Dictyonema* is within a 20 to 30 m-thick transitional zone above the base of the Dol-cyn-afon Member. The prevailing rock types in the member are dark grey mudstone and silty mudstone with rare tuffaceous mudstone forming laminae, lenses and beds at intervals throughout the member. The abundance of organic carbon, framboidal pyrite, and the abundance of trilobite exuviae in this formation suggest that, after the turbulent conditions during the deposition of the Ffestiniog Flags Formation, calm shallow conditions returned with short periods of emergence prior to uplift at the end of the Tremadoc epoch. PMA, AAJ

Biostratigraphy

The Mawddach Group ranges from the St David's into the Tremadoc epoch. The evidence relating to the biostratigraphy of the group is given in Allen, Jackson and Rushton (1981) and a summary only is given here. The zonal sequence is shown in Table 3.

Although the graptoloid *Dictyonema* is of significance in the Tremadoc the biostratigraphy depends in the main on the distribution of trilobites. Several typical fossils were illustrated by Allen and others (1981, pls. 16, 17), and those from beds around the Merioneth – Tremadoc boundary by Rushton (1982). Lake (1906, 1908, 1913, 1919 and 1935) figured some *Paradoxides*, agnostids and several olenids from the present area, but many of these are from Belt's collection, (held by the British Museum (Natural History) and BGS) and though well preserved are very inadequately localised. A selection of fossils is illustrated in Plates 4 and 5.

Clogau Formation

The presence of the *Tomagnostus fissus* Zone is indicated by records of trilobites such as *Tomagnostus fissus*, *Paradoxides hicksii* (Plate 4.8, 4.9) and *Eodiscus punctatus* s.l. from various localities in the valley of the Mawddach and in the Afon Llafar. The same zone is known from the lower part of the Nant-pig Mudstones at St Tudwal's Peninsula (Nicholas, 1915).

The *Hypagnostus parvifrons* Zone is recognised in the Llafar and Afon Prysor by the presence of *Ptychagnostus ciceroides*, *Peronopsis fallax depressa*, *Pleuroctenium granulatum* s.l., *E. punctatus* s.s. and *Hartshillina spinata* (Plate 4.6). The *parvifrons* Zone is also represented in the upper beds of the Nant-pig Mudstone. The fauna of the *Ptychagnostus punctuosus* Zone includes *Paradoxides davidis*, *Holocephalina primordialis*, *Meneviella*

Plate 2

1 Interbedded quartzose siltstone and dark grey mudstone at the base of the Maentwrog Formation at Aber Amfra harbour, Barmouth; folded and showing tectonic and sedimentary boudinage (L 1268).

2 Thinly interbedded quartzose siltstone and light grey fine siltstone of Ffestiniog Flags Formation on Precipice Walk. The thick unit is a microdiorite sill which terminates on the right (L 1277).

Plate 3
Photomicrographs of Cambrian sedimentary rocks

1 Greywacke from lower Maentwrog Formation showing vein of pyrite and some disseminated crystals. (E 47191, plane polarised light, ×25).

2 Quartz wacke from Gamlan Formation with quartz and feldspar grains in sericite-chlorite matrix. (E 47195, crossed polarisers, ×25).

3 Tuffitic mudstone in Gamlan Formation. (E 48305, crossed polarisers, ×25).

4 Lens of garnet-quartz in mudstone of Gamlan Formation. Above the dense garnet band in the upper part of the picture and at the bottom are chlorite-garnet bands. Two veinlets are chlorite. (E 49447, plane polarised light, ×25).

5 Mudstone in Clogau Formation with framboidal pyrite, vein pyrite and wispy carbonaceous material. (E 47193, plane polarised light, ×25)

6 Cleaved fragmental mudstone from Dolgellau Member (clear spots are holes), with goethite staining. (E 43141, plane polarised light, ×25).

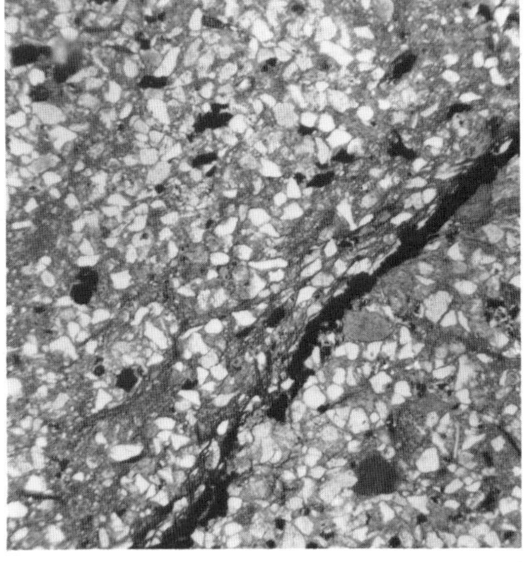

1

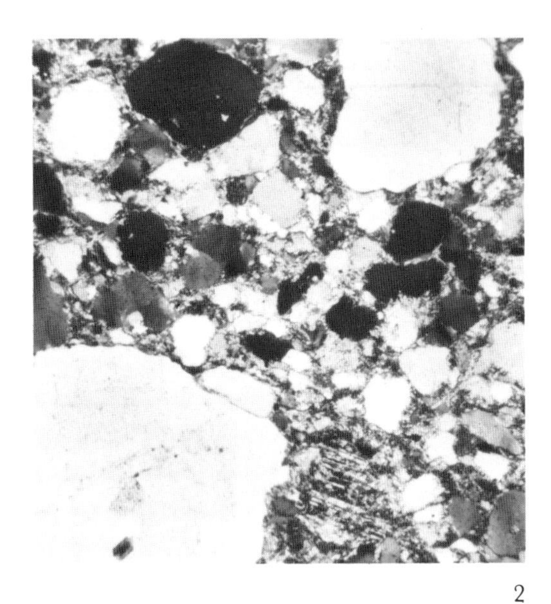

2

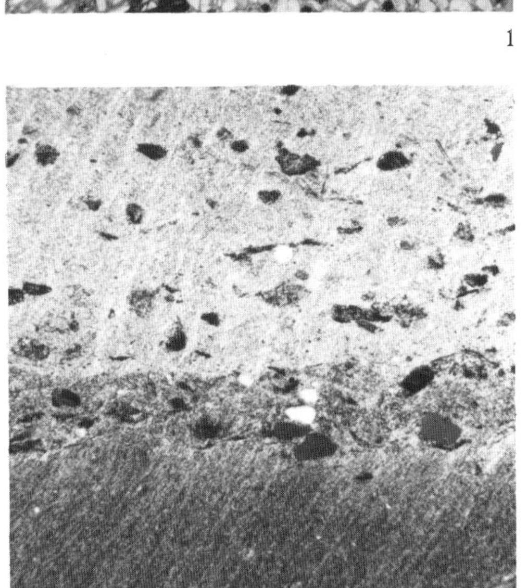

3

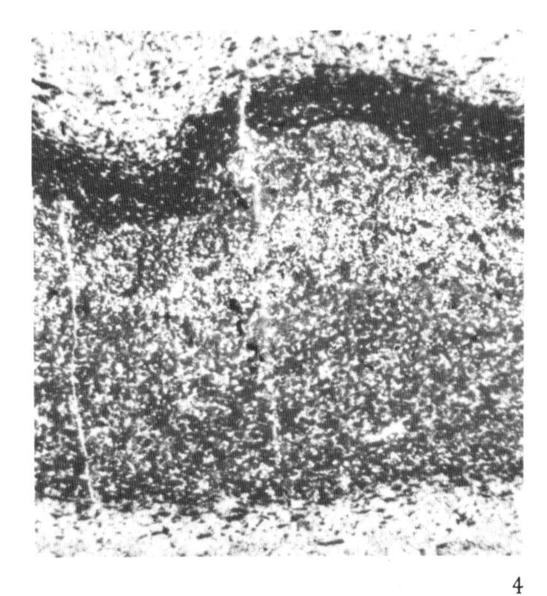

4

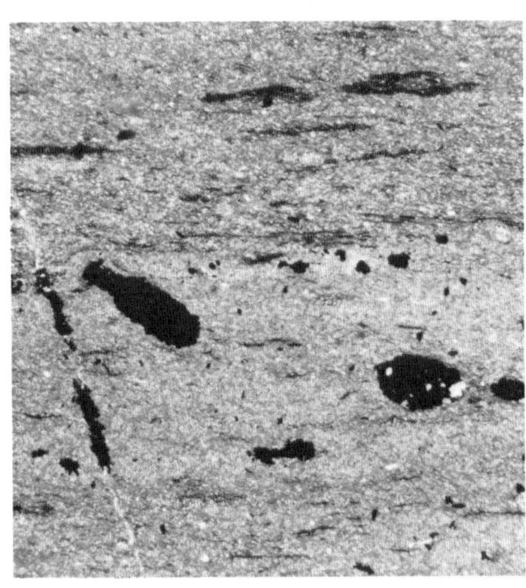

5

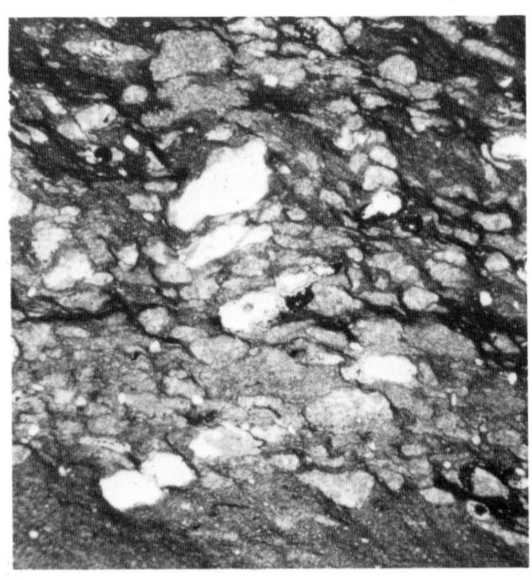

6

Table 3 Biostratigraphy of the Mawddach Group

LITHOSTRATIGRAPHIC DIVISION			ZONAL SUCCESSION		SUBZONES RECOGNISED HERE	SERIES/EPOCH
Mawddach Group	Cwmhesgen Formation	Dol-cyn-afon Member		*Dictyonema*	*D. f. flabelliforme* *D. f. sociale*	TREMADOC (part)
		Doigellau Member		*Acerocare*		MERIONETH
				Peltura scarabaeoides	*C. linnarssoni* *C. bisulcata*	
			NOT PROVED	*Peltura minor* *Protopeltura praecursor* *Leptoplastus*		
	Ffestiniog Flags Formation			*Parabolina spinulosa*	*P. spinulosa* *P. brevispina*	
			NOT PROVED			
	Maentwrog Formation			*Olenus*	*O. cataractes* *O. truncatus* *O. gibbosus*	
			NOT PROVED	*Agnostus pisiformis*		
			?	*Lejopyge laevigata*		
	?non-sequence		NOT PRESENT	*Solenopleura brachymetopa*		St DAVID'S (part)
	Clogau Formation			*Ptychagnostus punctuosus* *Hypagnostus parvifrons* *Tomagnostus fissus*		
Harlech Grits Group (part)	Gamlan Formation					

This table is reproduced by permission of Yorkshire Geological Society (P.Y.G.S. 1981, p. 306)

venulosa, *Anopolenus henrici* and agnostids, most commonly *Peronopsis scutalis* s.l. and *Ptychagnostus punctuosus* (Plate 4). This zone is the highest recognised in the Clogau Formation. Nicholas (1915, p. 103) showed that the *punctuosus* Zone is absent from the Nant-pig Mudstones.

The faunas of the Clogau Formation resemble closely those from the Menevian Beds of the St David's area, the Abbey Shales of the English Midlands and the Manuels Brook Formation of south-east Newfoundland and show some similarity to the faunas from the strata of the '*Paradoxides paradoxissimus*' stage in Sweden. The fauna was evidently widely distributed and is thought to represent the inhabitants of a broad marine shelf region of calm, probably cool, but not necessarily deep water; the prevalence of blind trilobites (agnostids, *Eodiscus*, *Meneviella*, *Holocephalina*, *Hartshillina*) outnumbering those with eyes (*Paradoxides*, solenopleurids) suggests that their environment was dimly lit, possibly because the water was turbid (cf. Clarkson, 1967).

Maentwrog Formation

The Maentwrog Formation ranges from the late St David's Series to the Merioneth Series (*Olenus* Zone). At one locality in the Afon Llafar agnostids of Middle Cambrian aspect suggest the presence of the *Lejopyge laevigata* Zone at the base of the Maentwrog Formation (Allen and others, 1981, p. 307).

The *Agnostus pisiformis* Zone is not proved but the greater part of the Maentwrog Formation is referable to the *Olenus* Zone and is characterised especially by *Homagnostus obesus* (Plate 4.5). The Subzones of *Olenus gibbosus* (with *Glyptagnostus reticulatus*) and *O. truncatus* occur in the lower half of the Maentwrog Formation and roughly the upper half of the formation may be referred to the Subzone of *O. cataractes*. In the east of the Harlech Dome *O. micrurus* is commoner than *O. cataractes*, but the two species appear to occur at about the same level in the formation.

Most of the trilobites were found in the interbeds of pyritous mudstone representing periods when the sea-floor conditions may have been oxygen-poor. During such periods the environment of deposition of the Maentwrog Formation presumably resembled that of the contemporaneous Outwoods Shales in the English Midlands (Taylor and Rushton, 1972, p. 12). The conditions during deposition of the arenaceous parts of the formation seem to have been unfavourable for the preservation of trilobites, though the presence of trace-fossils indicates the possibility that they were active on the sea-floor.

Ffestiniog Flags Formation

Rare *Homagnostus obesus* near the base of the Ffestiniog Flags shows that the upper part of the *Olenus* Zone extends into the formation. The presence of *Parabolinoides bucephalus* and a *Parabolina* fragment in the topmost beds shows that the upper part of the formation is referable to some part of the *P. spinulosa* Zone (Allen and others, 1981, p. 311). The greater part of the formation yields little apart from the brachiopod *Lingulella davisii* which is not biostratigraphically diagnostic. Thus the zonal classification of the beds around the boundary between *Olenus* and *P. spinulosa* zones is not clear in North Wales. A comparable uncertainty surrounds this interval in the much thinner sequence in the Nuneaton area (Taylor and Rushton, 1972, p. 21) and also in the Swedish standard succession, in which an unfossiliferous interval separates the beds with proved zonal fossils (Westergård, 1944, pp. 28–29, pls. 4–6).

The Ffestiniog Flags were deposited in shallow water in conditions which suited *Lingulella davisii* but were not so favourable to trilobites although their traces have been observed. At the time of the transition to the Cwmhesgen Formation the water was less turbulent and the conditions favoured smaller *Lingulella* and *Parabolinoides bucephalus*.

Cwmhesgen Formation

The Cwmhesgen Formation spans the Merioneth–Tremadoc series boundary, which is taken at the base of the beds with *Dictyonema flabelliforme* s.l., and coincides for practical purposes with the boundary between the Dolgellau Member and the Dol-cyn-afon Member.

The Dolgellau Member extends from the *P. spinulosa* Zone to the *Acerocare* Zone of the Scandinavian faunal sequence. The *P. spinulosa* Zone is well represented near the base of the member at several localities (Allen and others, 1981, p. 314) where *P. spinulosa* and *Orusia lenticularis* (Plate 5.20) abound, associated with rare specimens of *Parabolinoides bucephalus* and *Pseudagnostus sp*. The overlying zones with *Leptoplastus, Protopeltura praecursor* and *Peltura minor* are not proved in the east of the Harlech Dome, but the *Peltura scarabaeoides* Zone is represented in the neighbourhood of Foel Gron by rich faunas representing the lower two subzones. The *Ctenopyge bisulcata* Subzone has yielded the following trilobites: *Ctenopyge (Ct.) bisulcata, Ct. (Ct.) directa, Ct. (Ct.) falcifera, Ct. (Ct.) pecten, Lotagnostus trisectus, Micragnostus rudis, Parabolinella* aff. *caesa, Parabolinites? williamsonii, Peltura scarabaeoides scarabaeoides,* and *Sphaerophthalmus humilis*. Lake's (1913, pl. 10, fig. 9) specimen of *Ct. (Ct.) teretifrons* is associated with *S. humilis*; by comparison with Scandinavian occurrences its horizon is probably in the *Ctenopyge linnarssoni* Subzone, but the *bisulcata* Subzone remains a possibility.

Other assemblages which contain *Sphaerophthalmus major* may represent the overlying *linnarssoni* Subzone or a slightly higher horizon (Allen and others, 1981, p. 315), since *S. major* occurs above *S. humilis* in the Merevale No. 1 Borehole in the Nuneaton area (Taylor and Rushton, 1972, pl. 2). The following are known to occur in the Harlech Dome in association with *S. major*: *Lotagnostus trisectus, Micragnostus rudis, Parabolinella caesa, Parabolinites? longispinus, P.? williamsonii, Plicatolina* cf. *quadrata* (Plate 5.15) and *Peltura scarabaeoides westergaardi*.

The blocks with the type specimens of two other species, *Hedinaspis? expansa* (Henningsmoen, 1957, p. 210) and *Pseudagnostus obtusus* (Lake, 1946, p. 339) have no associated fauna so their horizons are uncertain; but *H.? expansa* at least came from the same general area, suggesting the *P. scarabaeoides* Zone.

These faunas are similar to those of the upper Dolgellau Member in the Tremadoc Anticline, the contemporaneous White-Leaved-Oak Shales of the Malvern Hills, the Monks Park Shales of the Nuneaton area, and to the *scarabaeoides* Zone faunas of Scandinavia. Henningsmoen (1957, pp. 79–82) discussed this type of specialised fauna, dominated by olenids and adapted to oxygen-poor conditions on the sea-floor.

Above the *scarabaeoides* Zone is a thickness of dark grey laminated mudstones referable to the *Acerocare* Zone. The fauna (Rushton, 1982), includes *Parabolina heres* accompanied by *Araiopleura stephani, Beltella nodifer, Parabolina angusta?, Parabolina frequens, Plicatolina kindlei* and *Shumardia alata*. The *Parabolinella sp.* recorded by Rushton (1982, pl. 3, figs. 23–25) is here tentatively identified with *P. contracta*, and some at least of Rushton's material of *Niobella homfrayi* may, on account of the comparatively narrow thoracic axis, be better referred to *N. h. preciosa* (Plate 5.10). Both *P. contracta* and *N. preciosa* were described from beds close to the base of Tremadoc correlatives in western Nei Monggol, North China (Lu, Zhou and Zhou, 1981). An example of the conodont *Cordylodus proavus* was collected near the base of the Zone and *Parabolina* cf. *acanthura* occurs near the top. There are species of non-olenid trilobite genera which anticipate

Plate 4 Fossils from the Maentwrog and Clogau formations.

1 *Olenus micrurus* Salter, latex cast of GSM 8948, ×2. *O. cataractes* Subzone, about 7 km E of Trawsfynydd (A. Selwyn coll.) Salters's original specimens; the lower one is lectotype (Henningsmoen, 1957, p. 107).

2 *Olentella rara* (Orlowski), Zs 840, ×1.5. *O. cataractes* Subzone, Roadside, Ffridd Dol-y-moch [7648 3298]. Pres. Mr C. Byrne.

3 *Olenus austriacus* Yang?, cranidium preserved in sandstone, British Museum (Natural History) 59287, ×3. *O. gibbosus* Subzone, Cefndeuddwr, Mawddach valley [near 730 261]. (J. Plant coll.).

5 *Homagnostus obesus* (Belt), lectotype, BM(NH) I.7646, ×6. *O. gibbosus* Subzone, Afon Mawddach, Dolmelynllyn [7295 2373]. (T. Belt coll.).

4, 7 *Ptychagnostus punctuosus* (Angelin), cephalon and pygidium, both ×4. *P. punctuosus* Zone, Afon Llafar [7342 3673]. RU 6441, 6440.

6 *Hartshillina spinata* (Illing), ×3. *H. parvifrons* Zone, waterfall in Afon Llafar [7357 3649]. RU 6702.

8, 9 *Paradoxides hicksii* Salter, presumably both from the *T. fissus* Zone. Fig 8. Flattened but typical cranidium, BM(NH) I.7729 (T. Belt coll.), ×2. 'Dolgelly' (locality uncertain). Fig. 9. Latex cast of GSM 10133, Salter's monotype specimen, ×1. A. Selwyn coll. from 'Cwm Eisen' (Cwmhesian), Mawddach valley [near 739 277].

10 *Meneviella venulosa* (Salter), latex cast of RU 6492, ×2. *H. parvifrons* Zone, above waterfall, Afon Llafar [7356 3657].

11 *Paradoxides davidis* (Salter), a small but typical specimen; BM(NH) 59269, ×1. Presumably *P. punctuosus* Zone, Tyddyngwladys, Mawddach valley [near 732 274]. (J. Plant coll.).

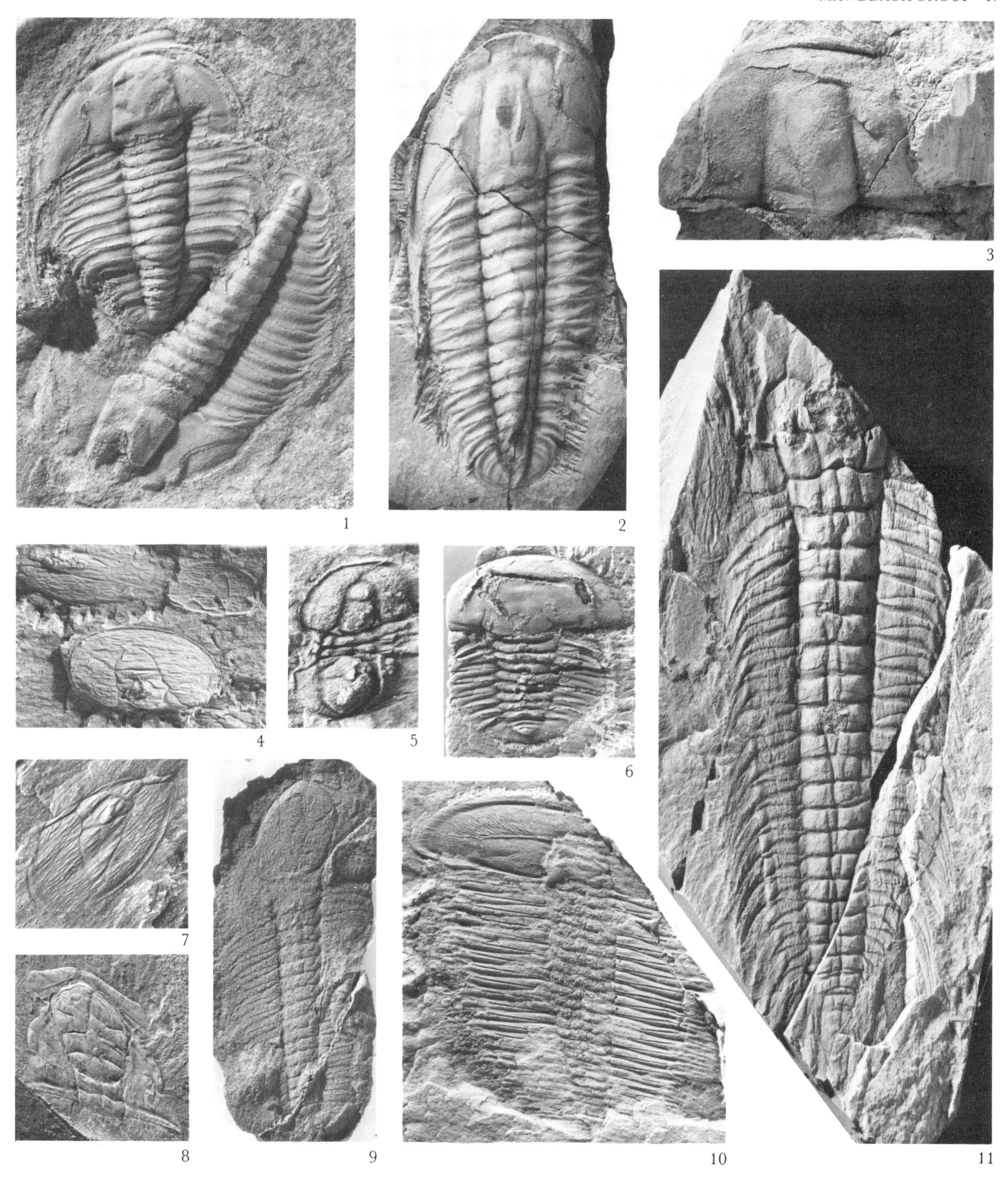
1
2
3
4
5
6
7
8
9
10
11

the faunas of the overlying Tremadoc Series, including *Acanthopleurella, Dichelepyge, Hysterolenus, Proteuloma, Pseudagnostus* (*Neognostus*) and *Psilocephalinella*. The appearance of these forms is presumably controlled in part by the change in environment which is reflected in the facies change from black to dark grey laminated mudstones.

The base of the Dol-cyn-afon Member is practically coincident with the appearance of dendroid graptolites of the *Dictyonema flabelliforme* group. The earliest *Dictyonema* are small forms with few dissepiments which are allied to *D. f. sociale* and *D. f. parabola* (Allen and others, 1981, p. 319); these occur just above a development of the *Acerocare* Zone in the forestry road at Bryn-llin-fawr, and sporadically elsewhere. They are taken to mark the base of a local *D. f. sociale* Subzone of the *flabelliforme* Zone. At slightly higher levels in the Dol-cyn-afon Member the *Dictyonema* include *D. f. belgicum* and synrhabdosomes of *D. f. sociale* (Plate 5.4; Allen and others, 1981, pp. 319–320). The shelly fauna associated with the *Dictyonema* in the lowest part of the member is less rich than that of the *Acerocare* Zone: most of trilobites of that Zone are absent, only *Beltella nodifer, N. homfrayi* subsp., *Psilocephalinella innotata* and *S. alata* persisting into the *sociale* Subzone. The trilobite *Boeckaspis hirsuta* was found just above the lowest *Dictyonema* in the road section at Bryn-llin-fawr, and the brachiopod *Eurytreta sabrinae* seems to become common only above the base of the *sociale* Subzone. The *sociale* Subzone appears to be over 100 m thick in the area of Bryn-llin-fawr.

The good continuity and correlative potential of the faunal and sedimentary record to the east of the Harlech Dome, especially the road section at Bryn-llin-fawr and the stream at Dol-cyn-afon, favour the definition of the base of the Tremadoc Series in that area (Rushton, 1982).

Above the *sociale* Subzone *D. flabelliforme flabelliforme* (*sensu* Bulman) is the predominant graptoloid. This and the appearance of *Boeckaspis mobergi* mark the base of the locally recognised *D. f. flabelliforme* Subzone. Rare specimens of the trilobite *Anacheirurus*? accompany the long-ranging *N. homfrayi homfrayi, P. innotata* and *S. alata*, but fossils are generally uncommon apart from small inarticulate brachiopods (*Lingulella spp., Eurytreta sabrinae*). However, at a quarry above Twr-y-maen, a larger fauna of the *flabelliforme* Subzone was collected: sponge spicules (stauractins), *Eurytreta sabrinae,* Bellerophontid, *'Hyolithes'* cf. *magnificus* (common), Agnostid (smooth), *Apatokephalus sp.*, *Micragnostus* cf. *bavaricus* (common) (Plate 5.1 & 5), *Niobella homfrayi smithi*, *Parabolina*? (fragment), *Platypeltoides sp.*, *Proteuloma* cf. *geinitzi* (common), *Shumardia curta*, Echinoderm (fragments, columnals?) and burrows.

Fossils are scarce from beds above those bearing *D. flabelliforme flabelliforme* in the district, and no higher zone in the Tremadoc Series is definitely recognised, although Wells (1925, p. 473) recorded fragments of *Asaphellus homfrayi* from a slate trial at Craig-fâch which, if correctly identified, suggests a horizon above the *flabelliforme* Zone. AWAR

RHOBELL VOLCANIC GROUP

The volcanic origin of the rocks in the area around Rhobell Fawr [787 257] was first determined by Ramsay (1881), and their stratigraphical position between the Tremadoc and Arenig was demonstrated by Cole and Holland (1890). Wells (1925) made the first detailed map and defined the Rhobell Volcanic Group (p. 532) as ' ... an extrusive phase, the products of which were essentially andesitic, and a phase of minor intrusion when sills of diorite-porphyry were injected'. Kokelaar (1977, 1979) remapped the Rhobell Fawr and redefined the group to exclude the intrusive rocks. As a result of detailed petrographical and chemical studies he also showed that the extrusive rocks are basaltic, not andesitic, in composition, and challenged earlier interpretations that the rocks are essentially tuffaceous (Ramsey, 1881; Wells, 1925).

The Rhobell Volcanic Group is now defined to contain all the volcanic and volcaniclastic rocks exposed on the south-eastern side of the Harlech Dome (Figure 5); it lies unconformably above sedimentary rocks of the Mawddach Group

Plate 5 Fossils in the Dol-cyn-afon Member (Tremadoc Series) and the Dolgellau Member (Merioneth Series).

1, 5 *Micragnostus* cf. *bavaricus* (Barrande), latex casts of cephalon and pygidium, RU 5599, 5598, both ×8. From *flabelliforme* Subzone, quarry above Twr-y-maen [7920 3125].

2 *Boeckaspis hirsuta* (Brögger), latex cast of cranidium, RX3,×6. Base of *sociale* Subzone, Bryn-llin-fawr forestry road [7905 3069].

3 *'Hyolithes'* cf. *magnificus* Bulman, large operculum, RU 9748, ×2. Horizon and locality as for Fig. 1.

4 *Dictyonema flabelliforme* (Eichwald) *sociale* (Salter), synrhabdosome, RU 6149, ×1. From *sociale* Subzone, Afon Melau [7947 2256].

6 *Proteuloma* cf. *geinitzi* (Barrande), latex cast of RU 4555A, ×4. Horizon and locality as for Fig. 1.

7, 8 *Linnarssonia belti* (Davidson), two of Davidson's syntypes, ×15. Fig. 7, brachial valve, is the lectotype selected by Cocks (1978, p. 23), British Museum (Natural History) BB 5809. Fig. 8, pedicle valve, is BM(NH) BB 5808. Lower Tremadoc, Craig-y-dinas [779 296 approx.].

9 *Eurytreta sabrinae* (Callaway), pedicle valve, latex cast of RU 6806, ×12. Track north of Ty-newydd-y-mynnydd [7996 2657].

10 *Niobella homfrayi* (Salter) *preciosa* Lu & Zhou, latex cast of RU 5377, ×1.5. *Acerocare* Zone, stream west of Dol-cyn-afon [7926 2881].

11, 12 *Parabolina heres* Brögger, both from *Acerocare* Zone. Fig. 11, cranidium HN 219, ×3, Nant yr Helyg [7983 2836]. Fig. 12, latex cast of pygidium RU 9154, ×2. Bryn-llin-fawr forestry road [7904 3068].

13 *Shumardia alata* Robison & Pantoja-Alor, latex cast of RU 5573, ×10. *Acerocare* Zone(?), Afon Cwmhesgen [7913 2990].

14 *Plicatolina kindlei* Shaw, cranidium RU 5176, ×8. *Acerocare* Zone, stream west of Dol-cyn-afon [7924 2878].

15 *Plicatolina* cf. *quadrata* Pokrovskaya, associated with fragments of *Pelura scarabaeoides westergaardi* Henningsmoen and *Sphaerophthalmus major* Lake, RU 4940, ×3. Stream, Rhobell-y-big [7882 2827].

16, 17 *Lotagnostus trisectus* (Salter), cephalon and pygidium, latex casts of RU 4741, 4773, both ×6. *P. scarabaeoides* Zone, *bisulcata* Subzone, stream on Rhobell-y-big [7859 2845].

18 *Peltura scarabaeoides scarabaeoides* (Wahlenberg), RU 4762, ×3. Horizon and locality as for Figs. 16, 17.

19 *Parabolinites? longispinus* (Belt), latex cast of GSM 49669, the counterpart of the lectotype BM(NH) I.7592, ×3. Collected by Belt from near Rhiw-felyn, exact locality unknown. It is associated with *Sphaerophthalmus major* and the horizon and locality may approximate to those of Fig. 15.

20 *Orusia lenticularis* (Wahlenberg), HN 56, ×2. Bryn-llin-fawr forestry track [7795 3017].

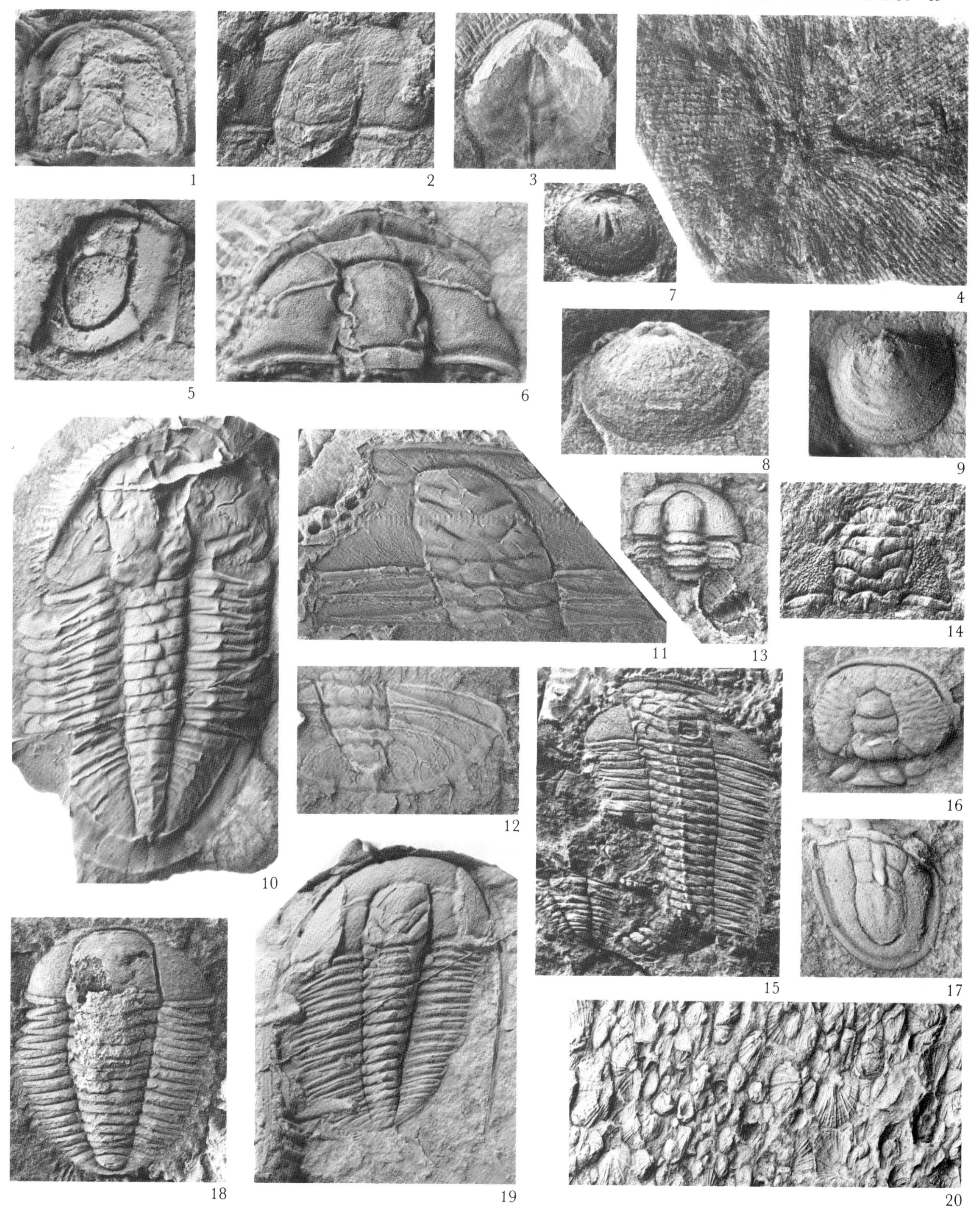
1
2
3
4
5
6
7
8
9
10
11
12
13
14
15
16
17
18
19
20

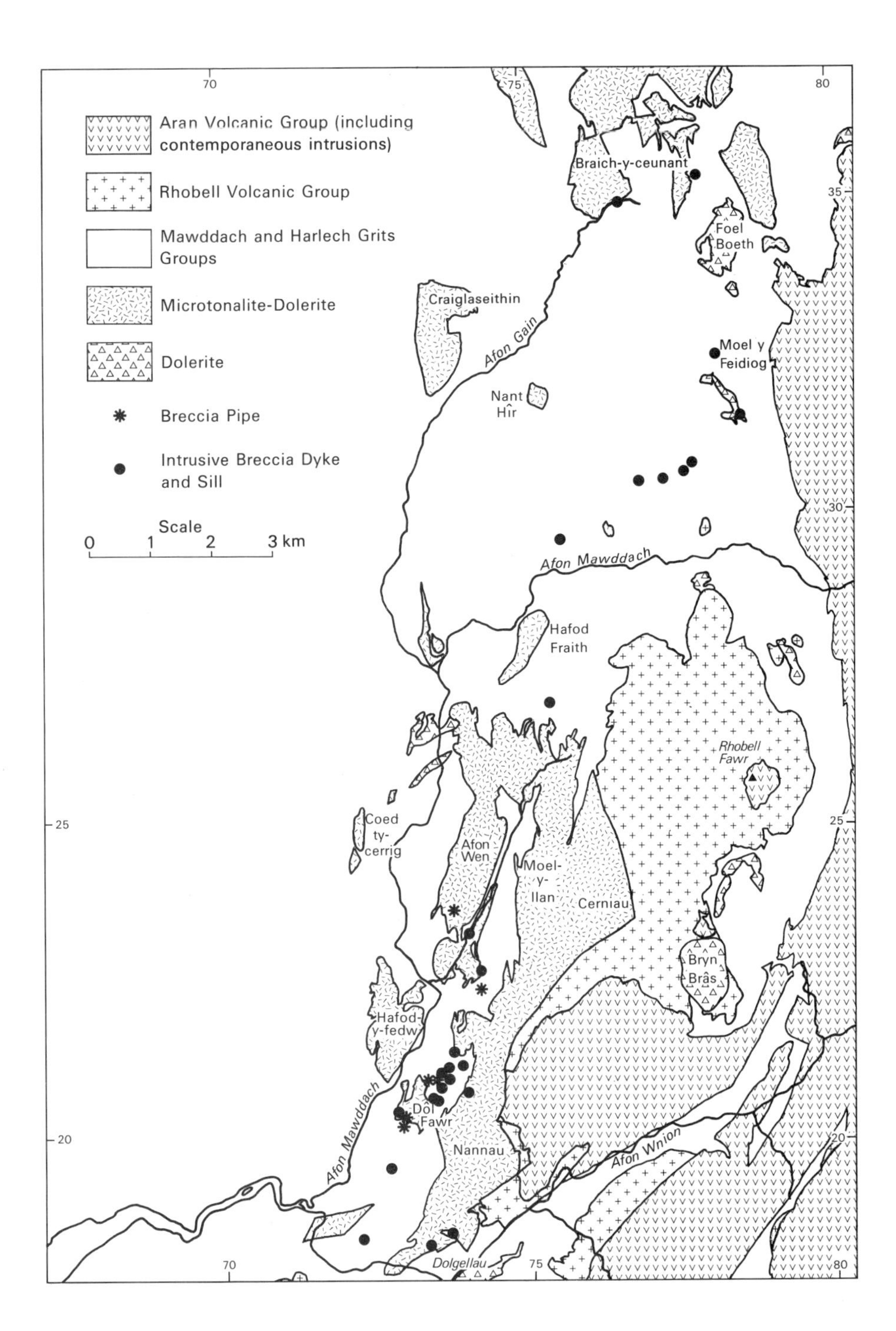

Figure 5 The Rhobell Volcanic Group and the major contemporaneous intrusions and intrusive complexes

and below the Allt Lŵyd Formation at the base of the Aran Volcanic Group. The main area of exposure of the group is the mountain of Rhobell Fawr, where Kokelaar (1979) estimated a thickness of 3.9 km of lavas has been erupted, but there are a number of outliers to the north, east and south. Together they comprise the eastern and southern parts of a single volcano. The effusive rocks were quietly erupted in a subaerial environment on to an eroded surface of locally folded Cambrian sedimentary rocks from a set of N-trending fissures represented by the Cerniau and Moel-y-llan intrusion complexes (p. 56), north of Llanfachreth. The unconformable relationships between the group and the Ffestiniog Flags and Cwmhesgen formations are well demonstrated around the main area of outcrop on Rhobell Fawr and in some of the outliers.

The rocks in all the areas are strikingly similar. They consist of mostly mixed compound and multiple basaltic flows (Walker, 1971) and variously reworked blocky autoclastic lava breccias. The amount of reworking increases upwards in the succession giving rise to talus breccias associated with local unconformities and coarse-grained, moderately well-bedded mass flow and alluvial deposits. Mainly it is these fragmental rocks that Kokelaar (1977) believed were misinterpreted as pyroclastic by earlier workers. Laharic (mud flow) breccias occur locally at the base. Acid extrusive rocks have not been identified with certainty. Bedded acid rocks of uncertain origin, and now highly altered, occur underneath the Allt Lŵyd Formation in an isolated area north-west of Foel Offrwm where they cannot be related to any other part of the group. Dacitic rocks here and in an outlier near Berth-Lŵyd [7425 1884] to the south are of uncertain origin.

According to Kokelaar (1977, 1979) the basalts show wide diversity in minor characteristics, but conform to two main rock types. The commonest is porphyritic with phenocrysts of augite and plagioclase; a less common variety, which occurs throughout the group, but is most abundant among the earliest effusions, contains large phenocrysts of pargasite and diopside or salite. Ultramafic cognate cumulate blocks occur within lavas and breccias of amphibole-bearing basalts. In his synthesis of the petrogenesis Kokelaar (*op. cit.*) included microdiorite and microtonalite from the comagmatic intrusive phase, and concluded that all these rocks were genetically linked by a process of crystal fractionation dominated by pargasite, the least fractionated rocks being the amphibole-bearing basalts. Using Ti-Y-Zr discriminant diagrams he classified the volcanic rocks as transitional between low-K tholeiites and calc-alkaline basalts.

Dewey (1969) first proposed that the Rhobell Volcanic Group was erupted during south-easterly subduction of ocean floor beneath Wales, and Kokelaar (1977) confirmed that the group shows the characteristics typical of island-arc, destructive plate margin volcanism. Kokelaar (1979) also related the N-trending folding which took place before, during and immediately after the Rhobell volcanism to repeated movement of fault blocks that were initiated during a phase of regional NE–SE compression related to the onset of subduction. It is Kokelaar's view that this phase of subduction must have ended with the Arenig transgression, to account for the subsequent change into a back-arc tensional environment, which Dunkley (1979) considers is appropriate for the Arenig–early Caradoc volcanism in the area. The start of the Rhobell subduction is less easy to define. Beds of reworked tuff are present in the upper parts of the Dolgellau Member (p. 14) and tuffaceous mudstone occurs throughout the overlying Dol-cyn-afon Member (p. 14). In the Migneint, Lynas (1973) recorded ashy beds low in the equivalent Afon Gam Formation. Prior to the volcanic episode indicated by these rocks the last evidence of volcanism in this part of Wales is in the Middle Cambrian Gamlan Formation (p. 12). It is likely, therefore, that the eruption of the Rhobell Volcanic Group marked the last stage in a period of subduction which began probably near the end of the Merioneth and lasted throughout the Tremadoc epoch.

K-Ar ages, determined by standard methods on five amphibole separates from the Rhobell Volcanic Group and one from comagmatic microtonalite, are reported by Beckinsale and Rundle (1980). The results range from 430 ± 14 to 475 ± 12 Ma, which they argue reflect loss of radiogenic ^{40}Ar. Thus, they accept the age of 475 ± 12 Ma as a minimum for the extrusion of the Rhobell Volcanic Group. More recently Kokelaar, Fitch and Hooker (1982) have published a K-Ar age of 508 ± 11 on pargasite phenocrysts from metabasalts in the group and suggest that the large (up to 4 cm long), virtually unaltered crystals are more likely to have resisted argon loss during regional metamorphism than partially altered or even smaller crystals. PMA, BPK

DETAILS OF ROCK TYPES

Augite- and plagioclase-phyric basalts

Augite- and plagioclase-phyric basalts comprise more than 99 per cent of the volcanic group at outcrop. The absolute and relative abundance of the phenocryst phases are highly variable and this factor, more than any other, accounts for the great variability in appearance of the lavas in the Rhobell Volcanic Group. The rocks weather generally grey or green, in places purple, and the phenocrysts or their pseudomorphs are commonly prominent. The lavas are non- to poorly vesicular. There is no systematic succession or lateral continuity of petrographical types.

Augite phenocrysts constitute from less than 1 to 5 per cent of the mode. They range from 0.2 to 5 mm in diameter, are anhedral, subhedral or euhedral and fresh or replaced. Normally zoned crystals and glomeroporphyritic clusters occur in places. Oscillatory zoning with concentric bands of minute inclusions are common near crystal margins. Plagioclase phenocrysts, which constitute 1 to 30 per cent of the mode, are usually subhedral, and range in size up to 4.5 mm. The crystal habit is variable and skeletal crystals are common. Normal zoning, oscillatory zoning and concentric bands of minute inclusions near crystal margins are common. Flow-alignment of lath-like crystals occurs and in places a pilotaxitic texture is developed between the larger phenocrysts. The composition ranges from oligoclase to andesine (An_{20-34}) and the mineral is assumed to be albitised. The groundmass is commonly very fine-grained and invariably altered; secondary minerals include chlorite, calcite, quartz, epidote, tremolite-actinolite, limonite and pyrite. Limonitic pseudomorphs after magnetite, 0.05 to 2 mm, are present in places.

Amphibole-bearing basalts

These basalts are characterised by phenocrysts of pargasite and diopside or salite with or without plagioclase (Plate 6). There is a negative correlation between the modal content of plagioclase and ferromagnesian phenocrysts. The amphibole generally predominates over the pyroxene in modal abundance but the relative proportions are variable. In most lavas pargasite phenocrysts (Table 4) constitute 1 to 3 per cent of the mode. In a few there may be up to 20 per cent, and in varieties in which cumulate blocks are abundant this value reaches 47 per cent. The crystals are commonly 5 to 15 mm long, though some reach 4 cm in length. They are mostly euhedral; angular fragments of crystals are common, and in some an edge cuts across oscillatory zoning suggesting late-stage fragmentation. Some phenocrysts have complex embayments containing groundmass and in others there are primary inclusions of actinolite. The pargasite is strongly pleochroic from colourless to light olive green or brown. Normal and oscillatory zoning and margins crowded with inclusions are not unusual. The diopside or salite phenocrysts commonly constitute 3 to 4 per cent of the mode but can reach 10 per cent. They commonly form colourless euhedra, up to 7 mm long, some of which are intensely cracked. Normal and oscillatory zoning occur. Andesine (An_{35-40}) phenocrysts, presumably albitised, constitute up to 20 per cent of the mode. They are on average 2 mm in length. Though commonly euhedral they are usually partially or completely replaced. The groundmass is invariably fine-grained, non-vesicular, and predominantly composed of plagioclase and chlorite with calcite, quartz, epidote and limonite.

Ultramafic cognate cumulate blocks

Cognate cumulate blocks occur sparsely in some amphibole-bearing basalts but are most abundant in breccias of such basalts in which the ferromagnesian phenocryst content ranges up to 50 per cent of

Plate 6

1 Porphyritic amphibole-bearing basalt from Rhobell Volcanic Group (L 1274).

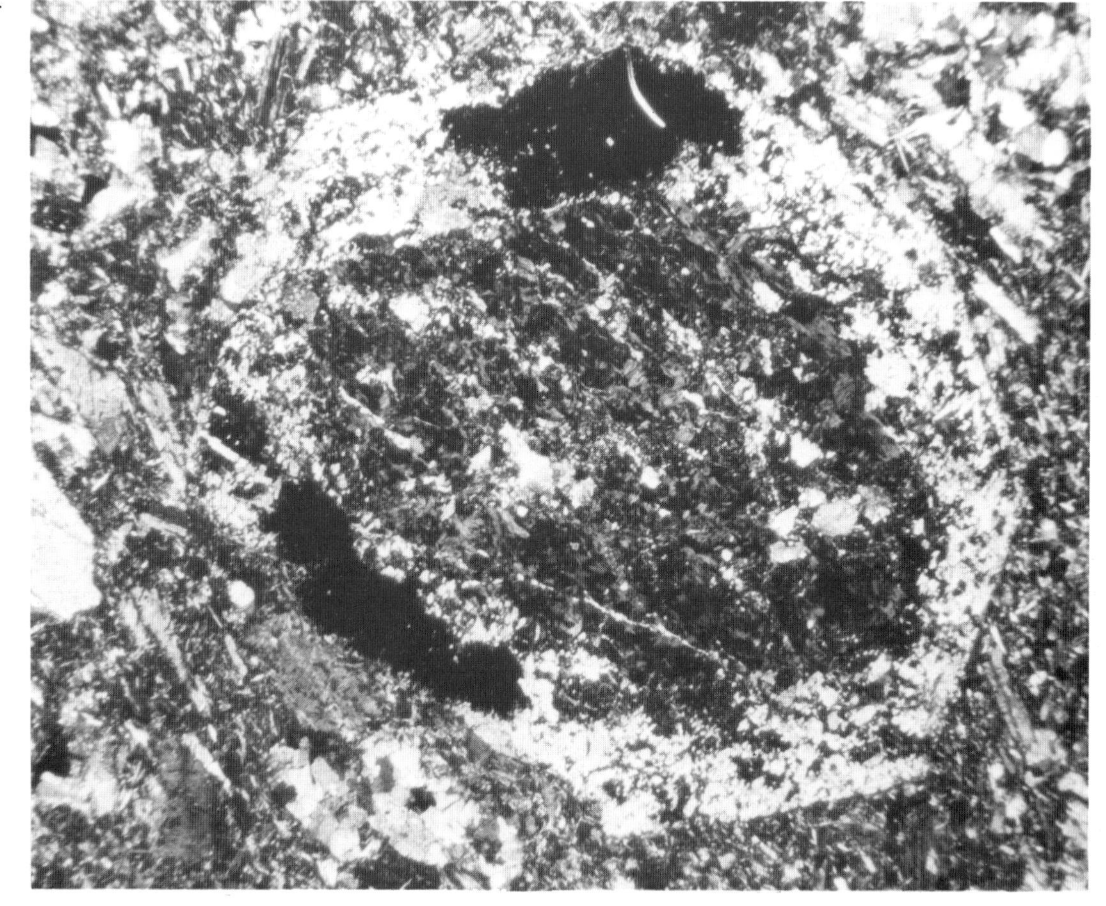

2 Pseudomorph after amphibole of chlorite and minor epidote with colourless amphibole preserved around the rim; from porphyritic basalt in Rhobell Volcanic Group (E 44744, crossed polarisers, ×40) (Neg. No. 14706).

the mode. Most blocks are rounded and some reach 16 cm in diameter. Many are partially altered but others are remarkably well preserved and layering has been found. Textural types include pargasite adcumulates, pargasite heteradcumulates, with primary actinolite as the poikilitic phase, and pargasite mesocumulates in which the trapped intercumulus melt is generally represented by secondary minerals, although in some blocks skeletal plagioclase is present. Cumulus diopside or salite is present in some blocks while titaniferous magnetite and sphene are generally present in small amounts.

Table 4 Chemical analyses of minerals in the Rhobell Volcanic Group

	1	2	3	4	5	6
SiO_2	40.20	41.20	40.17	55.62	50.77	49.27
TiO_2	1.65	1.57	1.96		0.38	0.40
Al_2O_3	13.25	14.35	15.50	1.04	5.42	5.92
Cr_2O_3					0.48	
Fe_2O_3	3.09	4.07				
FeO	7.73	7.34	9.47*	10.19*	5.41*	8.16
MnO	0.16	0.17				
MgO	15.60	15.30	14.75	17.24	15.40	13.71
CaO	11.00	9.77	12.13	13.33	23.01	23.20
Na_2O	2.00	2.03	2.33			0.30
K_2O	0.57	0.41	0.41			
H_2O^+	1.77	1.74				
F	0.08	0.05				
Total	97.10	98.00	96.72	97.42	100.87	100.96

1 Phenocrysts of pargasite in amphibole-basalt.
2 Single phenocryst of pargasite (4 cm) from amphibole basalt.
3 Pargasite from pargasite adcumulate.
4 Actinolite in pargasite mesocumulate.
5 Diopside phenocrysts in cumulo-phyric amphibole-basalt.
6 Salite in pargasite adcumulate.

Analyses 1 & 2 by XRF and wet chemical; 3 to 6 by electron microprobe.

Analyst B. P. Kokelaar * Total iron as FeO

Autoclastic lava flow-breccias

Blocky autoclastic lava flow-breccias occur throughout the volcanic group. The extent of their development and reworking are variable, but systematic with respect to time; a factor considered by Kokelaar (1977) to be attributable to variations in the rate of eruption.

They are very variable in character. They normally grade into massive lava [7790 2795 and 786 274] and occur at the base as well as at the top and sides of the flow-units, thus forming a carapace or envelope with a massive core. The breccias range from 2 cm to 20 m in thickness and may persist laterally up to about 500 m. They are normally unsorted and unbedded. In some breccias grading occurs in restricted poorly defined zones, and is attributable to variations in the brecciation process in different parts of the unit. Highly localised textural discontinuities, suggestive of bedding, probably represent minor avalanches that occurred during the advance of the lava.

The autoclastic flow breccias rarely contain exotic blocks and are believed to be derived by brecciation of a lava containing xenoliths. Pseudobreccias resulting from large-scale incorporation of xenoliths are common. The xenoliths are angular, almost invariably of locally derived basalts, and were probably incorporated when the lava broke through or flowed across earlier breccias. Clasts within clasts occur in places. The clasts are generally 1 to 30 cm in maximum diameter but range from sand size up to 2 m. The large ones are most commonly roughly equant and polyhedral although in some deposits they show signs of abrasion. Chilled margins are rare.

In general the term matrix is difficult to apply to the breccias because a complete range of grain sizes is commonly present. The role of the matrix as interstitial or supporting with respect to the large clasts depends on the history of formation of the breccia. Where there was continued flow of the breccia due to pushing or crushing by the flow, abrasion and comminution of the clasts took place thus increasing the proportion of fine-grained material so that eventually the large clasts changed from framework- to matrix-supported and became rounded.

Laharic breccias

Laharic breccias [e.g. 7771 2209] are present at the base of the pile. They are 1 to 2 m thick, unbedded, ungraded, very poorly sorted, heterolithic and of limited lateral persistence. They comprise angular to well rounded clasts set in a supporting mudstone matrix. Clasts of locally derived lavas are ubiquitous, ranging from isolated plagioclase fragments to blocks 20 cm in diameter. Rounded and lenticular clasts of mudstone or siltstone are commonly deformed, disrupted or have feather-edge margins, suggesting that they were unlithified at the time of incorporation. A clast of fine sandstone from the Ffestiniog Flags containing *Lingulella davisii* has also been collected. BPK

DETAILS OF EXPOSURE

Rhobell Fawr

The volcanic rocks overlie the Cwmhesgen Formation on the northern and eastern sides of the outcrop and rest on the older Ffestiniog Flags Formation along the western side. Here the sedimentary rocks dip steeply beneath the volcanic rocks, and about 200 m N of the ruined Hafotty-Hendre [7656 2743] the volcanic rocks overstep a small syncline. PMA

In the eastern parts [7956 2544] the angle between the evenly bedded volcanogenic sedimentary rocks and the closely subjacent unconformity surface is about 40°. High angles between these surfaces also occur in several other places across the outcrop and show that the volcanic pile built up against west- and north-facing topographical slopes. The pre-volcanic topography was also highly irregular; single lava flow-units on the southern flanks of Rhobell Fawr [7892 2482] can be seen to have ponded against a ridge of underlying sedimentary rock approximately 50 m high. South-east of Cyplau [775 273] numerous xenoliths, rafts and larger exposures of mudstone amongst the volcanic rocks attest to the local thinness of the lava pile, and again demonstrate substantial subvolcanic relief. In addition, in this area the strong electromagnetic anomaly which characterises the Dolgellau Member is detectable through the volcanic rocks.

Successive lava flows overlap towards the east or south-east on to the eroded basement. In the west, flow-brecciation of the early lava flow-units is uncommon and is generally restricted to a very thin outer carapace [e.g. 7700 2729]. However, higher in the sequence, flow-units show a general increase in brecciation so that around Rhobell Ganol [7847 2755], for example, flow brecciated units are common. Amongst successively younger lavas and their breccias there is evidence of an increase in the extent of their erosion and redistribution. Around Eglwys Rhobell [786 261] three intraformational unconformities have been recognised, and elsewhere [7864 2693] fragmental deposits are interpreted as talus breccias. The unconformities are well defined by steeply dipping, moderately well-bedded breccias. In general the individual beds, up to 10 m thick, are very poorly sorted, ungraded and commonly markedly heterolithic. They are wet debris flow deposits rather than 'explosion tuffs' as suggested by Wells (1925, p. 476). Laterally [7951 2592], sequences up to 6 m thick, of evenly bedded, poorly sorted feldspathic sandstones and grits represent flash flood deposits in minor alluvial fans that occurred near the periphery of the volcano. The uppermost lavas of the group, around Ffridd Graigwen [79 25], are similar to those of the initial effusions with the exception of two flow-units with well-developed blocky flow brecciated carapaces [7851 2481]. Around this locality there are fine examples of pseudobreccia, and a wide range of autoclastic breccia textures.

The variations through the volcanic pile are considered to reflect an initial relatively high rate of effusion that gradually decreased so that there was an increasing length of time between lava incursions during which erosion and the various modes of redistribution could occur. The increasing extent of lava brecciation reflects a decreasing rate of effusion for successive eruptions. The uppermost lavas show a return to a higher rate of effusion.

When eruption ceased the volcano was deeply eroded and folded before the basal sediments of the Aran Volcanic Group were deposited over the volcanic rocks. Outliers of Ordovician rocks, discovered by Cole and Holland (1890) and described by Wells (1925), remain resting on the Rhobell Volcanic Group in several areas. On the summit of Rhobell Fawr [788 256] and on Moel Corsy-garnedd [777 233] there are distinctive closely spaced joints parallel to the unconformity surface in the top 10 m of the Rhobell Volcanic Group, probably caused by the relief of vertical compression due to erosional unloading, prior to the Arenig transgression. Evidence of the post Rhobell, pre-Aran Volcanic Group folding is available on the summit outlier, where the Ordovician rocks are folded into a syncline plunging 8° towards 194° and in the underlying Rhobell Volcanic Group a syncline plunges 50° towards 184°.

BPK

Craig y Dinas

An outlier of the Rhobell Volcanic Group caps the hill at Craig y Dinas [779 297] where it rests on soft-weathering mudstone of the Dolgellau Member. The junction is not exposed. The Rhobell Volcanic Group is best exposed on the western face where it consists of massive basaltic lavas with intercalated basaltic breccias, similar to the autoclastic breccias on Rhobell Fawr. At the north end of the outlier the breccia occurs in bands 4 to 5 m thick alternating with massive lava. Southwards these bands appear to coalesce and then die out leaving one small lens. The breccia contains blocks of basalt up to 1.30 m long in a matrix consisting predominantly of feldspar crystals. The junction between the breccia and overlying massive lava is sharp and irregular.

AAJ

Bryn-y-Gwin

About 0.5 km W of Bryn-y-Gwin near Dolgellau, on the hillside above the Mawddach estuary, a small fault-bounded area of the Rhobell Volcanic Group unconformably oversteps the boundary between the Dolgellau and Dol-cyn-afon members of the Cwmhesgen Formation. The unconformity is exposed hereabouts [7077 1776]. The volcanic succession is composed of 1.2 m of crudely bedded muddy volcanogenic breccia beneath 60 m of lavas. The muddy breccias, presumably laharic in origin, are black and carbonaceous (E 45899) or grey (E 45898), and contain clasts of basalt up to 1 cm long, altered feldspar crystals and some mudstone fragments. One bed, 10 cm thick, is composed mostly of fragments of lava and feldspar all intensely altered to carbonate.

The lowermost of the overlying lavas (E 45894 – 5) are richly porphyritic, amygdaloidal, feldspar-phyric, highly altered basalts. Above them is one flow, 30 m thick (E 45893), of porphyritic basalt with conspicuous phenocrysts 2.5 mm or more long of albitised plagioclase and, less commonly, chlorite and calcite pseudomorphs after ferromagnesian minerals.

Berth-Lŵyd

About 1.5 km NE of Dolgellau and 400 m E of Berth-lŵyd [7415 1888] an outlier of volcanic rocks rests across the boundary between the Ffestiniog Flags Formation and the Dolgellau Member. The sedimentary rocks are disturbed near the contact [7417 1882], and in this area are intruded by a large intrusion complex related to the Rhobell magmatic episode.

The area is underlain mostly by grey or dark green porphyritic basalt with phenocrysts of feldspar and pseudomorphs after (?) amphibole. There are some microdiorite intrusions. A light grey porphyritic dacitic rock with quartz phenocrysts is present in the northern part of the area [7429 1902]. It is not known whether the rock is intrusive or extrusive though farther north such rocks are intrusive into the base of the pile.

Foel Offrwm

On the south-western side of Foel Offrwm near the fishponds in the Deer Park, a narrow strip of volcanic rocks lies between the Nannau intrusion complex on the west, containing a thin wedge of grey siltstone of the Ffestiniog Flags Formation [7462 1975], and early Ordovician sedimentary rocks on the east. The boundary between the intrusion complex and the extrusive rocks is imprecise, because of the close similarity between the rock types on either side of the boundary. The Ordovician rocks, however, are clearly seen at two localities [7476 2016 and 7483 1982] to rest unconformably on the lavas. All the rocks here are intensely altered to sericite and calcite. Nevertheless there are sufficient original textural features preserved to recognise the characteristic basalt with large amphibole phenocrysts among the rocks of the Rhobell Volcanic Group. Porphyritic dacite (E 46061) and quartz andesite (E 46065) are also present, but as at Berth-lŵyd there is no evidence to show whether it is of intrusive or extrusive origin.

North-west of Foel Offrwm an area of volcanic rocks, while not strictly forming a structural outlier, is isolated from the main area of outcrop of the group by the broad spread of thick boulder clay south of Llanfachreth. A small number of outcrops above the edge of the boulder clay and below the Ordovician rocks on the mountain consist of north-westward-dipping volcaniclastic rocks and minor dacite (?) lavas about 60 m thick. The bedded rocks are mainly pale grey or greenish grey, hard, splintery, intensely altered to calcite, sericite and chlorite, locally pyritised and with a good schistosity (E 45885 – 8). Using grain size criteria the rocks range from siltstone to sandstone and muddy sandstone. The larger clasts are mostly igneous in origin and include pseudomorphs after amphibole and feldspar, partly altered feldspar and, in some rocks, plentiful quartz. Some of the quartz occur in discrete grains either attached to or with enclosed feldspar crystals. The nature and origin of these rocks are difficult to determine, particularly with respect to the acid igneous component in them. Rocks of this composition have nowhere been recorded among the extrusive elements in the volcanic pile.

PMA

Brithdir

In the area west of Brithdir [762 188] along the south-eastern side of the Wnion valley there is a large, but very poorly exposed, outlier of the Rhobell Volcanic Group. All the rocks in this area dip steeply, and the basal unconformity is steeply dipping to vertical. Around Hendre Gyfeilliad [755 181] and about 2 km NE [around 775 197] the volcanic rocks overlie the Dolgellau Member, but between these two areas there is a progressive north-westwards overstep on to the Ffestiniog Flags Formation. A dolerite sill lies at the junction between the Rhobell Volcanic and Aran Volcanic groups along most of its length.

Rocks of the group, here about 550 m thick, are almost continuously exposed along Afon Clywedog [759 184]. The group consists predominantly of basaltic lavas and breccias, but to the south-west of the river, where the basal contact is exposed in a small outcrop [7546 1811] near Hendre Gyfeilliad, feldspathic volcaniclastic rocks of uncertain origin rest on dark grey mudstone of the Dolgellau Member. In the river itself there is a small intrusion of altered microtonalite lying between black mudstone and the lowest basaltic lava. The volcanic rocks are mostly green and purplish green porphyritic augite- and plagioclase-phyric basalts conforming to the types recorded on Rhobell Fawr. They are mostly altered and contain plentiful epidote veinlets, but in places between Afon Clywedog and Lletyrhys [7632 1904] fresh augite crystals up to 2 cm long have been observed. South-west of Afon Clywedog, in the lower parts of the group, porphyritic amphibole-bearing basalts are common.

The breccias are composed of poorly sorted subangular to subrounded fragments of basalt ranging in size from less than 1 cm to about half a metre. They are usually massive with interlocking fragments and little or no interstitial matrix, though stratified breccias with a matrix-supported fabric have been found. PND, PMA

South-east of the Bala Fault

Two small outliers of dark green, highly altered basaltic lavas are present on the south-east side of the Bala Fault, one near Tyn-y-ffridd-ddu [7980 1974] the other about 1.5 km SW [7891 1847]. In both places the volcanic rocks rest on grey silty mudstone of the Dol-cyn-afon Member. The small size of the outliers suggests that the group is considerably thinner here than around Brithdir about 3 km to the west. PND

CHAPTER 3

Ordovician rocks: the Aran Volcanic Group

Throughout the Cambrian, sediment accumulated in a trough on the north-east margin of the European continental plate separated from oceanic crust to the north-west by the Irish Sea ridge; volcanism in Wales was minimal and is represented by thin bands of ash-fall tuff erupted from a distant source. Convergence of oceanic and continental crust along a south-easterly dipping subduction zone on the south side of the Iapetus ocean (Phillips, Stillman and Murphy, 1976) initiated a period of tectonic and volcanic activity in the late Cambrian which persisted into the Caradoc and in the Silurian led to closure of the basin. The initial eruption of basaltic lavas of the Rhobell Volcanic Group during the Tremadoc was followed in the Ordovician by a mixed suite of basalts, rhyolites, volcaniclastic and sedimentary rocks, which around the Harlech Dome constitute the Aran Volcanic Group.

Late Tremadoc folding and faulting resulted in a regional uplift which was greatest in the west. This initiated a period of intensive subaerial erosion in the Harlech district, but to the north sedimentation may have continued uninterrupted in local basins (Lynas, 1973). The Rhobell Volcanic Group was erupted subaerially on to an irregular topography (Kokelaar, 1979). A renewed period of folding was followed by further erosion reducing the Rhobell Volcanic Group and the hinterland of folded Cambrian sediments to a peneplain.

Shallow marine sediments of Arenig age rest on the Rhobell Volcanic Group and on the Cwmhesgen Formation in the Harlech district and overstep progressively older rocks to the west to rest on lower Cambrian and Precambrian rocks on the Llyn peninsula.

The base of the Aran Volcanic Group (Figure 6) is marked locally by the Garth Grit, the lowest member of the **Allt Lŵyd Formation**. This is a mature sandstone consisting predominantly of quartz grains, possibly derived by erosion of the Harlech Grits Group to the west, and deposited as a beach or sublittoral deposit. A period of relative volcanic quiescence followed the marine incursion, and the bioturbated siltstones and sandstones which overlie the Garth Grit

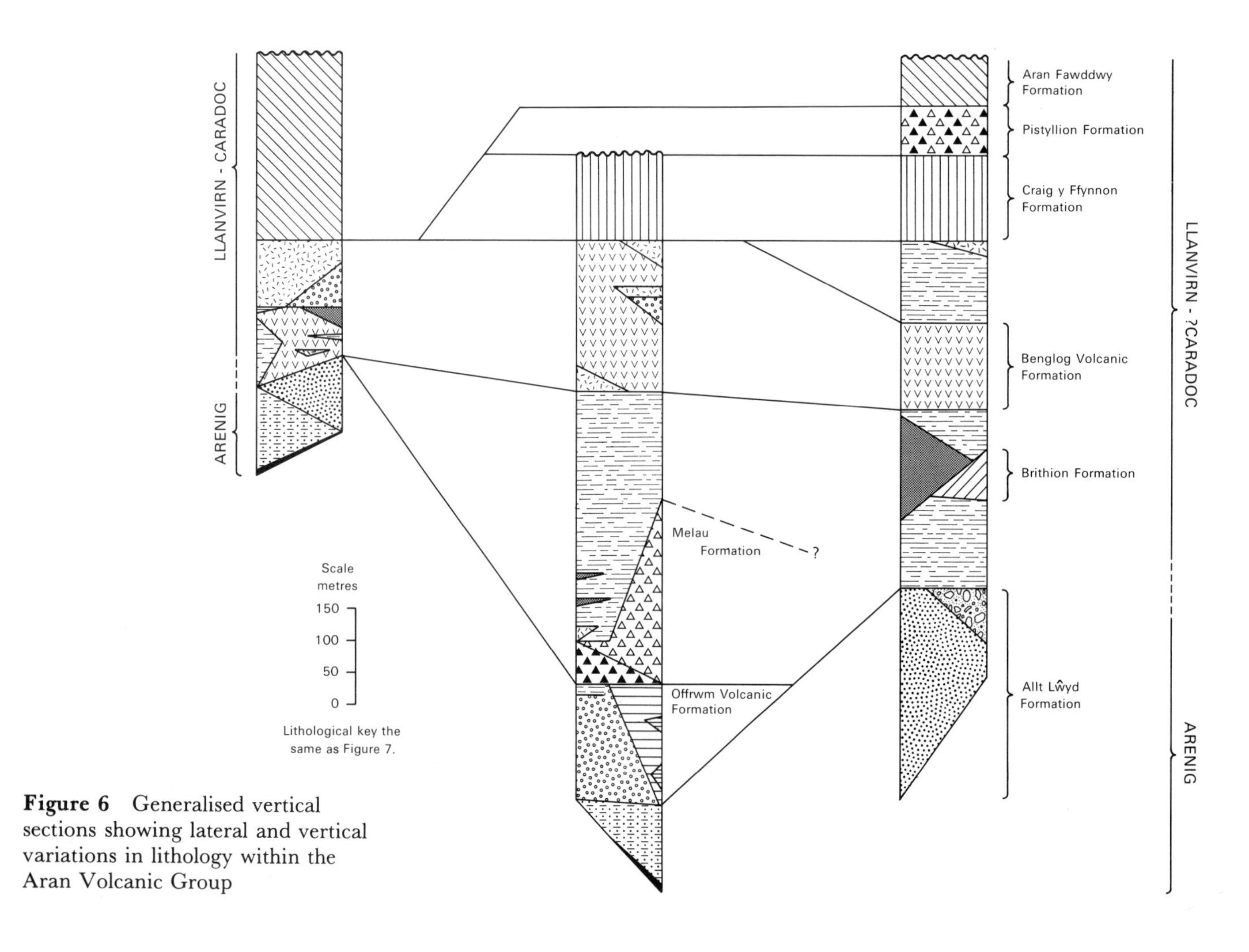

Figure 6 Generalised vertical sections showing lateral and vertical variations in lithology within the Aran Volcanic Group

are interpreted as mudflat deposits. Seismic instability is reflected in intercalated slumped beds and minor small scale penecontemporaneous faults.

Volcanic sandstones were introduced gradually from the south. Over most of the district they consist of recrystallised or partially recrystallised feldspar clasts generally assumed to have been derived from the Rhobell Volcanic Group. Southwards less altered lithic volcaniclastic rocks (tuffaceous sandstone, tuffite and tuff) predominate, particularly in the upper part of the Allt Lŵyd Formation. These consist of clasts of intermediate feldspar-phyric lava or intrusive material.

The Aran Boulder Bed is the topmost member of the Allt Lŵyd Formation to the south of the Bala Fault, and also consists of clasts of feldspar-phyric lava. It is best developed east of the district where it oversteps the lower part of the Allt Lŵyd Formation to rest on the Mawddach Group. The Boulder Bed has been interpreted as an alluvial fan; the lithic volcaniclastics to the north of the Bala Fault, although fine-grained and only slightly conglomeratic by comparison, are interpreted as the lateral equivalent. The lithic volcaniclastics show cyclic bedding sequences as well as thick graded and massive beds which could be interpreted partly as fluvial deposits and partly as debris flow deposits. The clasts contained in these lithologies are similar to the lavas of the Rhobell Volcanic Group but the volume of material and their uniformity preclude the known outcrops of the Rhobell Group as a source. It seems more likely that the lithic volcaniclastic rocks and the Boulder Bed were derived from a contemporaneous intermediate shallow intrusion or extrusive dome of similar composition to the Rhobell lavas, and that this source lay to the south-east of the present outcrop in the Aran Mountains. Ridgway (1971) commented on an apparent coarsening upwards of the crystals within the boulders and suggested that this may reflect the un-roofing of a lava dome, with the finer grained, more rapidly cooled, cover of the dome being eroded first followed by the coarser grained interior. The sequence is interpreted as alluvial fan gravels and sediment gravity flows deposited on a shallow shelf probably mainly in subaqueous conditions although to the south, near the source, the deposits may have been subaerial. The more altered feldspathic sandstone may represent the more distal deposits which have been further reworked.

The first evidence of acid volcanicity in the Ordovician is at the top of the Allt Lŵyd Formation where a local bed of tuffite consists of fragments of rhyolitic tuff or lava. Overlying the Allt Lŵyd Formation a thin siltstone contains scattered volcanic quartz and feldspar phenoclasts, showing that volcanic activity had subsided but had not ceased completely.

After the deposition of the Allt Lŵyd Formation a marked difference in environment developed between the north and south of the district (Figure 7). To the south a mixed suite of acid and basic extrusive rocks is interbedded with volcaniclastic sediments. This includes the Offrwm, Melau and Brithion formations. In the north these formations are absent and the Benglog Volcanic Formation rests directly on the Allt Lŵyd Formation. The gradual subsidence to the south was controlled by NE-trending faults subparallel with the Bala lineament, and the northern edge of the basin lay between Cae'r-defaid and Cwm yr Allt-Lŵyd. Within the district this part of the sequence is largely submarine although some of the eruptions producing the volcanic debris may have been subaerial. The Offrwm and the Brithion Formations are regarded as the lateral equivalents of a more complete sequence of volcanic rocks preserved to the south in Cader Idris and the Aran Mountains.

The first major acid eruption is marked by the **Offrwm Formation**. In the type area this consists of two flow units each with a thick massive component and a thinly bedded upper part. The bedding suggests that the pyroclastic debris was emplaced by turbidity flow, although it is not possible to say whether this was primary volcanic material ejected from a vent and carried as a submarine pyroclastic flow or secondary material recycled from unconsolidated tephra deposits. Some evidence of welding in the lower tuff suggests that this is a primary pyroclastic flow. At Cae'r-defaid this formation is represented by only a few metres of reworked banded ash.

The **Melau Formation** is the first of many basaltic eruptions. Around the Afon Melau, where it is thickest, it consists of tuff and tuffite intercalated with basaltic lavas and hyaloclastic flows and is likely to be near the vent. Elsewhere the basaltic rocks at this horizon are thickly bedded and unsorted, suggesting deposition from subaqueous gravity flows. On Foel Offrwm the tuffite is capped by pillow lavas. Higher in the sequence basaltic tuffite and mono-lithologic breccias are commonly interbedded with the siltstone and acid volcanic rocks, reflecting the discontinuous eruption of basaltic lava brecciated either by quenching in contact with water or by phreatic explosions. Temporary cones of brecciated lava may have accumulated around local vents but, unless protected by more resistant lavas, the unstable debris collapsed at intervals giving rise to numerous sediment gravity flows. In the south where the Melau Formation is absent the Offrwm Formation is overlain by a sequence of siltstone containing thick gravity-flow breccias composed of poorly sorted polymict volcanic clasts in a siltstone matrix.

The overlying **Brithion Formation** is associated with a rhyolitic lava dome to the east on 1:50 000 Geological Sheet 136 (Bala), but is only poorly represented within the district where it consists of a basal breccia overlain by welded ash-flow tuff. Dunkley (1978) has shown that the top of this tuff is truncated by an erosional surface presenting evidence of temporary emergence.

The Brithion Formation is overlain by a further sequence of siltstone and tuffaceous siltstone with a local thick accumulation of basic tuff on the southern edge of the district.

Uniform conditions were again established across the area with the deposition of the **Benglog Volcanic Formation**, and this probably involved a marine transgression north of Cwm yr Allt-Lŵyd. It is not clear if this part of the district was emergent during the deposition of the sediments and volcanic rocks to the south or if the upper part of the Allt Lŵyd Formation is a condensed sequence. The Benglog Volcanic Formation consists mainly of crystal tuff, but throughout the area it contains intercalated tuffaceous siltstone, pillow lavas (Plate 7.1) and hyaloclastite indicating subaqueous conditions.

The formation as a whole shows little evidence of systematic thinning but to the south intercalated hyaloclastite, lavas and gravity flow breccias are more com-

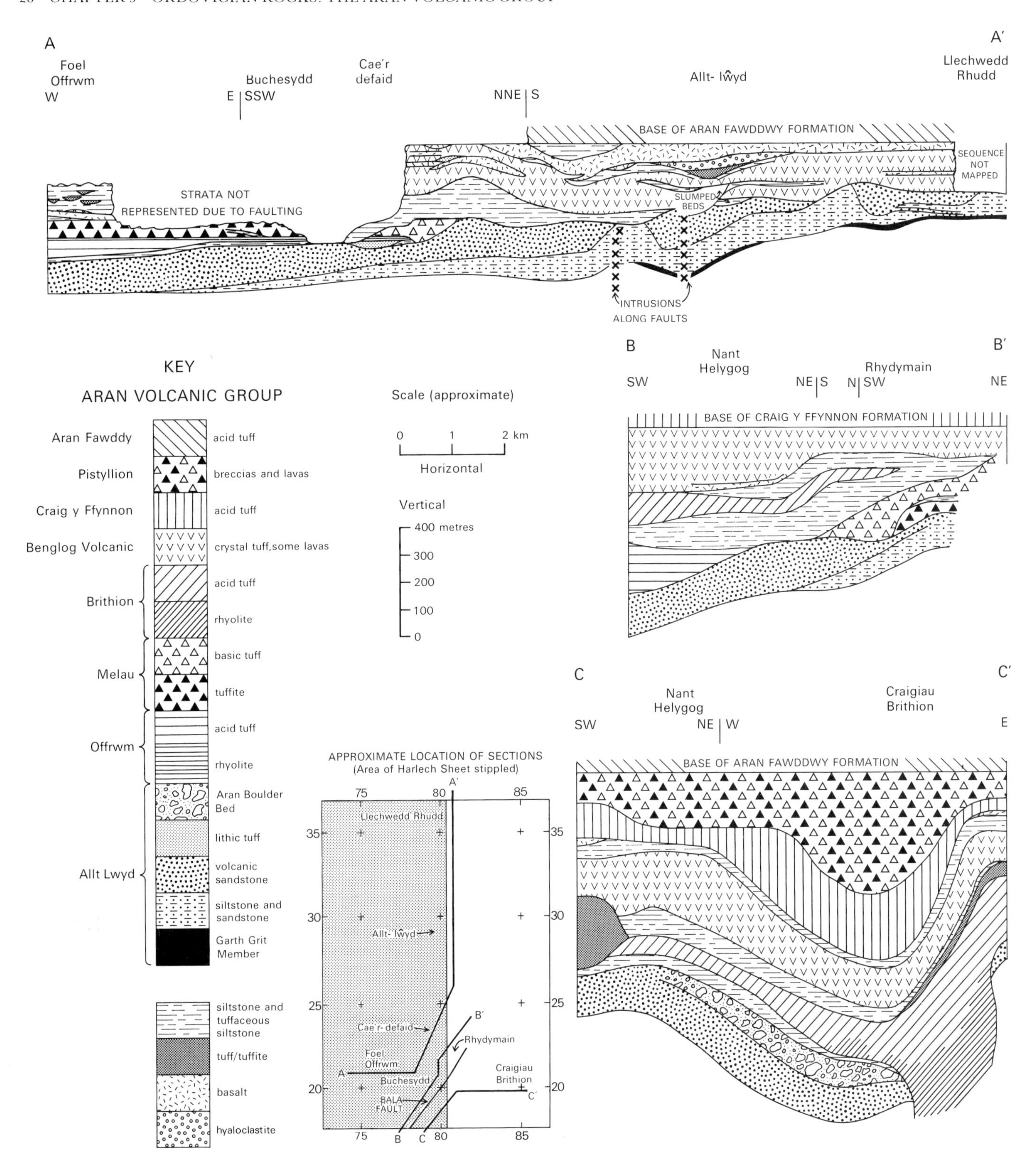

Figure 7 Variations in thickness of the formations in the Aran Volcanic Group on the eastern edge of Harlech and adjacent parts of Bala sheets

Plate 7

1 Pillow basalt in the Aran Volcanic Group (L 1301).

2 Columnar jointing in a dolerite sill within the Aran Volcanic Group (L 1289).

mon suggesting that the crystal tuff may be thinner than in the north. It is clear from texture and composition that the lavas and crystal tuff originated from different parent magmas. The hyaloclastite and some dolerite intrusions are intimately associated with the pillow lavas, and originated in a non-porphyritic magma which was erupted intermittently from a number of vents at different times.

The crystal tuff is grain supported, thickly bedded with poorly defined stratification and is composed mainly of fragmented or rounded euhedral crystals of oligoclase-andesine composition. The upper part of the sequence is more thinly bedded and contains fragments of pumice and rare but widely scattered euhedral quartz phenoclasts testifying to derivation from an increasingly differentiated magma. The feldspar phenoclasts are similar in composition to the feldspar-phyric clasts of the Aran Boulder Bed in the Allt Lŵyd Formation but it is unlikely that the crystal tuff had a similar source. To the north of the district, Lynas (1973) described quartz-latite consisting of feldspar phenocrysts in a devitrified groundmass, and blocks of this rock are incorporated in the tuff. The quartz-latite is intruded into and below the level of the Serw Formation, the stratigraphical equivalent of the Benglog Volcanic Formation. A partially crystallised magma of similar composition to these high level intrusions could produce crystal segregation during sustained subaerial explosive eruption where rapid release of pressure results in vesiculation and comminution of the molten fraction. Winnowing of the dust and pumice fragments from the eruption column may have resulted in a concentration of crystal clasts settling on the slopes of the volcano or into water. Seismic activity caused repeated sediment gravity flows of the crystal debris accumulated on the volcanic slopes and could have carried material into the basin as turbidity currents producing the apparently uniform and widespread crystal tuff.

The oligoclase-andesine bearing tuff does not occur above this horizon and the deposition of mudstone is interrupted by the acid ash-flow tuffs of the **Craig y Ffynnon Formation** and the products of the basaltic and andesite eruptions in the **Pistyllion Formation**, usually represented by laharic breccias or peperites. Both these formations wedge out to the east and north of the district.

The **Aran Fawddwy Formation** is the ultimate expression of the Ordovician volcanic activity. This thick deposit of

Table 5 Comparison of lithostratigraphic terms used within the Aran Volcanic Group

Fearnsides, 1905	Wells, 1925	Cox, 1925	Cox and Wells, 1927	Harlech Sheet (135)
Caradoc Upper Ashes of Arenig	Upper Acid Group	Craig-y-Llam Group	Upper Acid Volcanic Group	Aran Fawddwy Formation
Llandeilo Daerfawr Shales	Mudstones	Llyn-Cau Mudstones	Llyn Cau Mudstones	Pistyllion Formation
				Craig y Ffynnon formation
Lower Ashes and Agglomerates	Basic Volcanic Group	Pen-y-Gader Group	Upper Basic Volcanic Group	Benglog Volcanic Formation
		Llyn-y-Gader Mudstones and Ashes	Llyn-y-Gader Mudstones	AVG* with unnamed basic tuff horizon at base
		Llyn-y-Gadr Spilitic Group	Lower Basic Volcanic Group	
Llanvirn Olchfa or bifidus-shales	Shales (Zones of *D. bifidus* to *N. gracilis*)	Cefn-hir Ashes	Cefn-hir Ashes	Brithion Formation
		Crogenen Slates	Crogenen Slates	AVG* undivided
		Bryn Brith Grits	Bryn Brith Grits	
		Moelyn Slates	Moelyn Slates	
				Melau Formation
Arenig Filltirgerig or Hirundo-Beds Erwent or Ogygia-Limestone Henllan or Calymene-Ashes Llyfnant or Extensus-Flags Basal Grit	Lower Acid Group	Mynydd-y-Gader Group	Lower Acid Volcanic Group	Offrwm Volcanic Formation
	Shales		Pont King's Slate	AVG* undivided
	Basement Group	Basement Beds	Basement Group	Allt Lŵyd Formation (Garth Grit at base)

* AVG = undivided siltstones, mudstones, usually with interbedded volcanic horizons

acid ash-flow tuff can be traced from north to south of the district. In the Aran Mountains, Ridgway (1971) identified three cooling units and described welding. To the east of the district, on Geological Sheet 136 (Bala), the tuff is largely unwelded and contains at least three intercalations of sediment and basaltic laharic breccia suggesting submarine deposition.

Previous work

The earliest references to these volcanic rocks were by Sedgwick (1852) and Ramsay (1866, 1881). Fearnsides (1905) established a detailed stratigraphy of the Arenig area and his map, at a scale of 3 inches to one mile, covered the north-east of the district. The area south of Cwm yr Allt-Lŵyd to the Wnion was described by Wells (1925) and the stratigraphy was correlated with Cader Idris (Cox and Wells, 1927). The stratigraphic nomenclature used by these authors is compared with the present study in Table 5.

More recently Ridgway (1975) updated the lithostratigraphical nomenclature and defined the Aran Volcanic Group and its constituent formations. Broadly the formational names (Table 6) were based on those used by Cox and Wells (1915, 1921). Ridgway described the lithological and petrographical variations within the Aran Volcanic Group between Cader Idris and Arenig. However, the present work has questioned the reliability of these correlations. In particular, Ridgway accepted the suggestion of Cox and Wells (1927, p. 283) that the Cefn-hîr Ashes of Cader Idris are equivalent to the Platy Ashes and Agglomerates described by Fearnsides on Arenig Fawr. The present mapping has shown that these Platy Ashes and Agglomerates are clearly equivalent to the Basic Volcanic Group of Wells (1925) or the Upper Basic Volcanic (or Pen-y-Gader) Group of Cox and Wells (1927), here named the Benglog Volcanic Formation. To avoid further confusion in correlation, the formational names introduced by Ridgway (1975) have not been used, but his group name has been retained.

The ground to the south-east of the Wnion is poorly exposed and has been mapped by Dunkley (1978) as part of a larger project in the Aran Mountains. South of Cae'r-defaid, the volcanic and sedimentary lithologies become thicker and considerably more diverse than those which crop out to the north, with the Offrwm, Melau, Brithion, Craig y Ffynnon and Pistyllion formations wedging out northwards. Because of the variation in thickness of the lithologies, Cox (1925, p. 561) concluded that folding and faulting were already in operation during the deposition of the strata, a conclusion which is confirmed here.

There is some debate over the correlation of the volcanic strata, for while Dunkley (1978) agrees with the correlation of the Aran succession with that north of the Wnion (Table 7) he prefers to equate the Basic Volcanic Member of his Braich-y-Ceunant Formation (Benglog Volcanic Formation) with the unit termed the Lower Basic Volcanic Group (Cox and Wells, 1927) or the Llyn-y-Gafr Spilitic Group (Cox, 1925), and to equate the Pistyllion Formation with the Upper Basic or Pen-y-Gader Group. Dunkley argues that the geochemistry of the lavas and hyaloclastics of the Pistyllion Formation is distinctive and that the high silica andesites-dacites (60 to 63 per cent SiO_2) are the most fractionated rocks of the tholeiitic suite seen anywhere in the district. He compares them with high silica andesites-dacites analysed by Davies (1959) in the Upper Basic Group or Cader Idris. However, this is not a unique horizon, as Dunkley has also mapped andesite at a lower horizon [7882 1910] and comparative geochemical data are incomplete. Ridgway (1971), who mapped the Bwlch Oerddrws area (Table 6) which joins the Arans and Cader Idris Mountains, supports the lithological correlation made by Cox and Wells (1927).

The palaeontological evidence is meagre and not conclusive over most of the area (Table 5). Graptolites suggesting an upper Arenig age are recorded from a slate band in the Mynydd-y-Gader Group on Cader Idris (Cox, 1925, p. 547) and this group is equated with the Offrwm Volcanic

Table 6 Comparison of formation names used by Ridgway in the Bwlch Oerddrws area [800 192] with those depicted on the Harlech Sheet (135)

Harlech Sheet		Ridgway, 1971, 1975	
Formation	Lithology	Formation	Lithology
Aran Fawddwy	acid tuff	Craig-y-Llam	Ignimbrite
Pistyllion	mudflow breccias with intercalated tuffs and lavas	Craig-y-Bwlch	Mudflow
Craig y Ffynnon	acid tuff		Ignimbrite
Benglog Volcanic	intermediate crystal tuff with lavas hyaloclastites and minor siltstone	Pen-y-Gader	Mudflow
			Basic tuffs lavas and siltstones
AVG	siltstone and tuffaceous siltstone with unnamed basic tuff horizon	Nant Ffridd Fawr	Ignimbrite
		Llyn-y-Gadr	Mudflow
Brithion	acid tuff	Cefn-hir	Mudflow
AVG	siltstone and tuffaceous siltstone	Gwynant	Mudstone
Allt Lŵyd	Aran Boulder Bed	Pared-yr-Ychain	Boulder Bed
	Volcanic Sandstone		Siltstone, sandstone and feldspathic wackes
	Interbedded siltstone and sandstone		

The Offrwm and Melau formations are absent in this area.

Table 7 Comparison of lithostratigraphic names used to the north of the Harlech district

Fearnsides, 1905	Lynas, 1973		Zalasiewicz, 1981 and 1984		Harlech Sheet (135)
Upper Ashes of Arenig	Llyn Conwy Formation		Llyn Conwy Formation		Aran Fawddwy Formation
					Pistyllion Formation, Craig y Ffynnon Formation (present only in the south)
Daerfawr Shales	Serw Formation		Serw Formation		Benglog Volcanic Formation
Lower Ashes and Agglomerates					Brithion, Melau and Offrwm formations together with intervening siltstones (present only in the south)
Olchfa or Bifidus Shales					
Filltirgerig or Hirundo Beds					
Erwent Limestone	Carnedd	Siltstone Member	Henllan Ash Member	Carnedd	Allt Lŵyd Formation (Garth Grit Member at base)
Henllan Ash	Iago	Henllan Ash		Iago	
Llyfnant Flags	Formation	Siltstone	Llyfnant Member	Formation	
Basal Grit		Garth Grit	Garth Grit Member		

Formation. On Foel Offrwm this formation is overlain by basic rocks which have been tentatively equated with the Melau Formation outcropping in the Afon Melau to the east, and a record of *Merlinia selwynii* from this basic tuff indicates an early Arenig age. Thus these basic rocks are older than the Lower Basic Volcanic Group on Cader Idris.

A Llanvirn age is suggested for the Cefn-hir Ashes (Brithion Formation) and the Lower Basic Volcanic Group on the occurrence of *Didymograptus* '*bifidus*' in the underlying Crogenen Slates (Cox and Wells, 1927, p. 282) and *Glyptograptus teretiusculus* in the overlying beds (Llyn-y-Gader Mudstones).

East of Cae'r-defaid a thick sequence of siltstone which rests partly on the Melau Formation and partly on the Allt Lŵyd Formation is overlain by the Benglog Volcanic Formation and contains fossils of Llanvirn age. In addition, Wells (1925, p. 506) recorded a fauna of Llandeilo/Caradoc aspect *Nemagraptus gracilis* Zone) from these siltstones, but this has not been confirmed in the present study. On Cader Idris the Llyn-y-Gader Mudstones, lying between the Upper and Lower Basic Volcanic groups, has yielded the graptolite *Glyptograptus teretiusculus* which is Llandeilo in age, and suggests that the Benglog Volcanic Formation must equate with the Upper Basic Volcanic Group. The Llanvirn siltstone thins abruptly to the north so that the Benglog Volcanic Formation rests on the Allt Lŵyd Formation, overstepping the lower formations.

To the north, work by Lynas (1973) and more recently

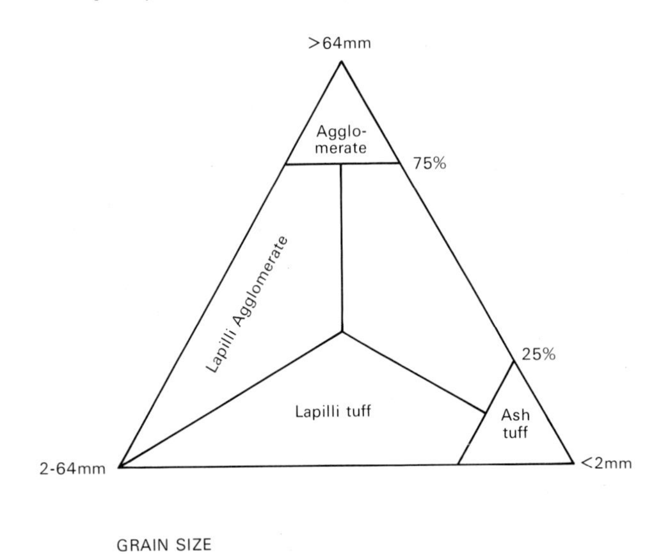

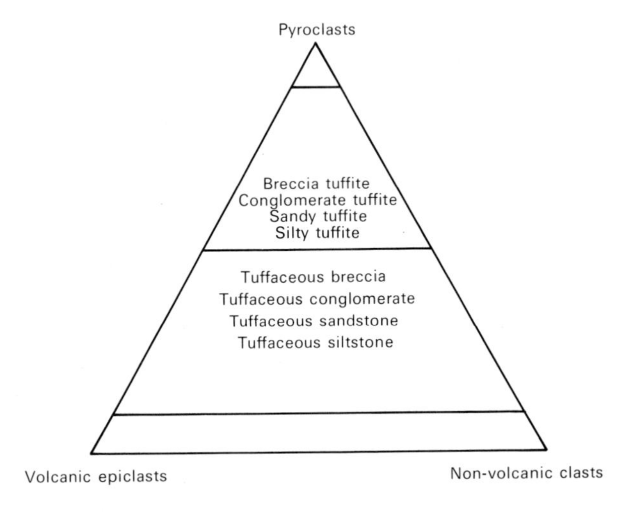

Figure 8 Classification of volcanic and volcaniclastic rocks

Zalasiewicz (1981, 1984 and *in press*)) has clarified the biostratigraphy in an area, part of which was previously mapped by Fearnsides (1905). The correlation of their work with that used on the Harlech Sheet is shown in Table 7. Zalasiewicz has shown that the Daerfawr Shales (included in the upper part of the Serw Formation), which Fearnsides assigned to the *Didymograptus murchisoni* Zone, are younger, containing a fauna indicating the *N. gracilis* Zone. He found that the palaeontological evidence suggests the presence of unconformities or non-sequences at the top and base of the Carnedd Iago Formation, (equivalent to the Allt Lŵyd Formation) which supports the lithological evidence in this district.

Nomenclature in this work

The volcanic rocks dominate the sequence and define the formations. Where the volcanic rocks are absent the interbedded siltstone and tuffaceous siltstone cannot be separated and are designated to the Aran Volcanic Group undivided.

The classification of the volcanic and volcaniclastic rocks is adapted from the interim report of the I.U.G.S. Commission on systematics of Igneous Rocks (Figure 8).

The sedimentary rocks are classified according to Pettijohn (1957) with the modification (Krumbein and Sloss, 1963) that rocks containing less than 15 per cent rock fragments and feldspar and more than 15 per cent detrital matrix are termed quartz wackes.

ALLT LŴYD FORMATION

This formation is equivalent to the Basement Group of Cox and Wells (1927). Over most of the outcrop the formation varies between 120 and 260 m in thickness, but to the south-east of the Bala Fault it is between 70 and 330 m thick, wedging out to the north-east of the district on the slopes of Aran Fawddwy (1:50 000 Geological Sheet 136: Bala).

The formation comprises four distinct sedimentary facies which interdigitate with each other but are broadly divisible stratigraphically into:

4 Conglomerate (Aran Boulder Bed Member)
3 Volcanic sandstone (tuffaceous sandstone and tuffite)
2 Interbedded siltstone and sandstone
1 Quartzose sandstone (Garth Grit Member)

Minor lithologies include oolitic mudstone and slump breccias.

The Garth Grit Member (Jennings and Williams, 1891) at the base of the Allt Lŵyd Formation corresponds to the Basal Conglomerate of Cox and Wells (1927). Though of wide lateral extent it is rarely more than 5 m thick. It consists of relatively well sorted, coarse-grained, mature quartzose sandstone with the composition mainly of protoquartzite, but ranging through to subgreywacke. The sandstone was laid down as a beach or sublittoral deposit on the eroded, emerged surface of upper Cambrian and Tremadoc rocks at the start of the Arenig. Its composition suggests derivation from a distant source, as the subjacent strata consist of silt, clay and volcanic material. On the Lleyn peninsula the Garth Grit overlies the Harlech Grits Group (Nicholas, 1915) which is a likely source of the coarse grains of clastic quartz.

Phosphatic nodules which occur within the Garth Grit were originally described as a ceramoporoid bryozoan, *Bolopora undosa* (Lewis, 1926). However, Hofmann (1975) interpreted these structures as oncolitic accretions and compared them to chemogenic geyserite stromatolites (Walter, 1972).

To the south the Garth Grit is thin and impersistent but has been recorded south of the Bala Fault and is present on the slopes of Rhobell Fawr and Foel Offrwm. It is exposed at Allt Lŵyd, the type-section for the formation, and north of the Mawddach it can be traced more or less continuously maintaining a thickness of 4 to 5 m. Locally, for example at Blaen Lliw [801 345], two thick beds of quartzose sandstone occur, separated by up to 4.5 m of banded siltstone containing laminae and thin beds of quartzose sandstone.

The interbedded siltstone and sandstone which overlies the Garth Grit Member forms a unit which corresponds approximately to the Llyfnant Flags of Fearnsides (1905) and to the Arenig Flags of Cox and Wells (1927). The beds are distinctive, with alternating beds and laminae of dark grey siltstone and light grey, fine-grained sandstone. The sandstone beds are generally bioturbated and some bedding planes on the siltstones are imprinted with trace fossils though other fossils are rare. The thickness of the unit changes abruptly across certain ENE- to NE-trending faults, which may have been active as growth faults during sedimentation, reflecting the tectonic instability of the volcanic terrain. In general the beds are poorly developed or absent in the south and reach their maximum thickness in the north. Around Moel Llyfnant they pass laterally into volcanic sandstone, but crop out again to the north of the district in the Migneint area where Lynas (1973) described them as streaky mudstone and siltstone.

Typically this unit consists of well bedded flaggy pyritous siltstone and pale grey sandstone in laminae and thin beds up to about 2 cm thick, but in places thicker sandstone beds, 30 to 40 cm thick, occur. There is much internal variation in the proportion of siltstone and sandstone. Where the beds are not disturbed cross-lamination is common, occurring as single lenses or as simple and multiple sets. The sandstone beds commonly show small-scale load structures at the base and sand injection into the overlying laminae.

In general the sedimentary structures have been disturbed to varying degrees by bioturbation. This takes the form of simple vertical and horizontal burrows which have subsequently filled with sand. Where activity is more intense the sedimentary fabric is disrupted and contorted (E 42165). In detail an analogy could be drawn between this lithology and the recent mud-flat deposits and sand flat deposits described by Evans (1965, pls. 17, 19, 20).

Some of the thicker sandstone beds, up to 2 m, show evidence of slumping especially where the sandstone contains silty laminae which may be folded or completely disrupted. Some of the beds have been cut by vertical burrows after deformation.

The volcanic sandstone unit is equivalent to the Henllan Ash and Erwent Limestone of Fearnsides (1905). It forms a complementary wedge to the underlying interbedded siltstone and sandstone, forming a thick continuous sheet in the south where the latter is thinnest. In the north the volcanic sandstone is intercalated with the interbedded siltstone and sandstone and shows a considerable variation in thickness. It is likely that the volcanic sandstones are diachronous, and erosion may have produced a local disconformity between the two units.

The term volcanic sandstone is used here to include tuffite, tuffaceous sandstone and epigenetic volcanic sandstone. The composition varies from sandstones, consisting mainly of highly altered feldspar clasts, to those consisting mainly of lithic clasts. The uniformity of the clasts seen in the lithic sandstones and tuffites seems to indicate a derivation from a contemporaneous volcanic source rather than from the erosion of the Rhobell Volcanic Group as has been previously suggested. Most clasts are andesitic or basaltic, but sparse conspicuous quartz phenoclasts in some beds and some acid tuffite beds attest to limited acidic eruptive activity.

Beds average 20 cm but are up to 2 m thick. The thicker units are conglomeratic in part. Cross and parallel stratification are common throughout, occurring as single sets or multiple cosets which are separated by massive or parallel stratified sandstone or by stringers of micro cross-stratification (Picard and High, 1973). Trough cross-stratification (Harms and Fahnestock, 1965) appears to be the most common form with some tabular (Figure 9) and low angle cross-stratification.

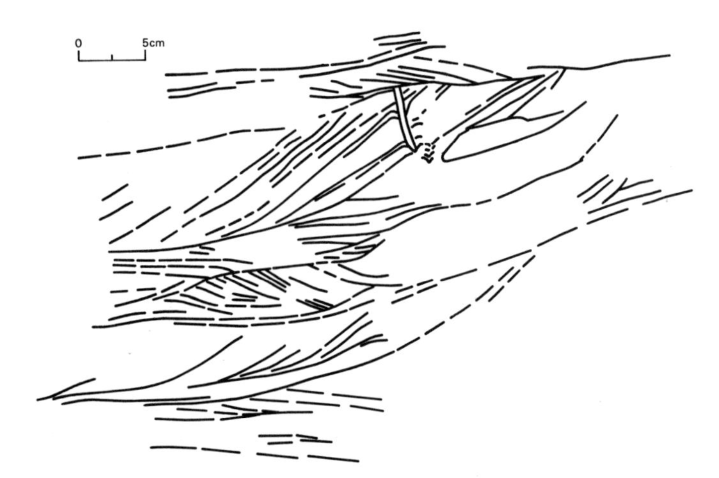

Figure 9 Tabular cross-stratification cut by vertical burrows in volcanic sandstone of Allt Lŵyd Formation

Numerous different associations of these bedding forms can be seen; a typical sequence consists of trough cross-stratification with pebbly laminae in the lower part passing up into parallel stratification with low angle cross-stratification and then into massive sandstone.

Coarsening upwards sequences can be seen particularly within the tuffitic horizons near Coedwig [7749 2152] and Cae'r-defaid [7971 2370]. Massive sandy tuffite, 1 to 2 m in thickness, passes up into interbedded conglomerate and cross-bedded tuffite which, in turn, is overlain by conglomeratic tuffite 1 to 2 m thick. AAJ

The Aran Boulder Bed Member is the highest unit in the Allt Lŵyd Formation on the south-east side of the Bala Fault. Its thickest development, 300 m, is to the east (Sheet 136). Within the district it is no more than 70 m thick on the eastern margin and can be traced only a kilometre southwestwards, where it is thinnest and finest and is represented by 20 m of fine matrix-supported conglomerate. It is composed of breccia and conglomerate. PND

Ridgway (1971, p. 149) described the boulders as consisting of a porphyritic felsite and observes that the size of the feldspar phenocrysts in the boulders increases upwards. From this he suggested derivation from a shallow intrusion or dome with a hydrothermally altered upper margin, the margin being the first part to be eroded. Elsewhere in his thesis he interpreted the boulder bed as alluvial fans deposited near a fault scarp. Cox and Lewis (1945, p. 73) suggested the Foel Ddu intrusion as a source for the clasts, but Dunkley (1978) considered that this intrusion is comagmatic with the younger Brithion Formation and could not, therefore, be a source of the Boulder Bed. Dunkley (*op. cit.*) interpreted the Boulder Bed as being derived from a positive area to the east of the Arans, which was probably fault controlled and composed of volcanic rocks similar to those of the Rhobell Volcanic Group.

Kokelaar (1979) also supported the view that the clasts within the Allt Lŵyd Formation could have been derived from the erosion of the Rhobell Volcanic Group, and suggested that the acid tuff (Offrwm Volcanic Formation) resting on top of the Allt Lŵyd Formation represents the end member of the Rhobell volcanic episode. Thus, it is possible that lavas similar in composition to the Rhobell Volcanic Group continued to be extruded during the Arenig period, possibly as domes as well as explosive material, contributing to the lithic tuffs and tuffites. The uniform composition of these beds and of the Aran Boulder Bed suggests a minimal mixing of lithic material from different sources such as would occur if it had been transported for any distance over variable terrain. It is, therefore, concluded that the Boulder Bed could have been derived from later eruptions of the Rhobell type, the source of which lay in the eastern part of the Aran Mountains, and that most of these rocks were reworked and deposited in a high energy environment such as an alluvial fan with associated debris flow.

DETAILS

Type section

The base of the formation is exposed in a tributary of the Afon Mawddach [7969 2912] which flows past the abandoned farm house of Allt Lŵyd. This is taken as the type section for the formation. The lowest 70 to 80 m are exposed in this tributary and the overlying beds are less completely exposed on the plateau above Ffridd yr Allt Lŵyd. The succession, with approximate thickness in metres, from top to bottom is as follows:

Allt Lŵyd Formation (Arenig)	
Calcareous breccia, angular fragments of siltstone, sandstone and basic igneous rock in calcite (reaches a maximum thickness of about 23 m and dies out within less than 1 km)	1.6
Volcanic sandstone, mid grey, coarse-grained with interbedded fine-grained sandstone and siltstone. Well bedded, cross-stratified sandstone alternates with bioturbated sandstone. Exposure poor	105–108

No exposure	48
Interbedded bioturbated sandstone with laminated and cross-laminated sandstone and siltstone. Bioturbated beds consist of medium- to coarse-grained sandstone with wisps and balls of siltstone, abundant burrows infilled with sandstone, beds 40 to 50 cm thick (E 50429)	42–45
Interbedded dark grey siltstone with thinly bedded sandstone and siltstone and with thicker sandstone beds (up to 1 m) scattered throughout. Sandstone is mainly feldspathic subgreywacke with some beds of quartz-dominated subgreywacke (E 50428). Bedding averages about 1 cm in thickness and the sandstone shows small-scale cross and parallel lamination, minor scour and fill structures. Bioturbation is common and small penecontemporaneous faults are present at some levels	100
Sandstone, brownish weathering with thin silty laminae	0.9
No exposure	4.5
Garth Grit Member: Sandstone, coarse-grained, massive with scattered pebbles to 2 cm long, quartzose, angular to subrounded, dark chloritic groundmass. Finer grained with thinner beds in upper part (E 50427)	4.12
CWMHESGEN FORMATION (Tremadoc)	
Dol-cyn-afon Member: Siltstone, dark grey	-

Quartzose sandstone (Garth Grit Member)

The sandstone is mid-grey in colour, thickly bedded, massive, or with poorly defined cross-stratification, or more rarely normal grading. Large-scale undulations occur on the upper surface. Locally the sandstone includes lenses up to 30 cm thick of quartz pebble gravel. The sandstone comprises angular to subrounded grains averaging 1 to 2 mm in diameter but with scattered pebbles up to 3 cm long. Sorting is moderately good and the sediment is grain-supported.

Quartz and quartzite grains make up to 86 per cent of the sediment. Quartz is clear or contains very small inclusions, possibly of rutile. Quartzite clasts are fine to coarse-grained and might have been derived from a recrystallised sediment or acid tuff. Lithic clasts constitute up to 15 per cent of the rock in the Allt Lŵyd area, but elsewhere the percentage of these clasts drops to about 4 per cent. They are mostly siltstone or igneous rock. Most of the igneous clasts consist of secondary chlorite, stilpnomelane and muscovite, but some are porphyritic and contain quartz phenocrysts and pseudomorphs of recrystallised quartz in a groundmass of altered lath-shaped feldspar and interstitial chlorite. Amphibole crystals are pseudomorphed either by stilpnomelane and chlorite or quartz and iron oxides. The matrix is generally less than 15 per cent, and consists predominantly of sericite with minor chlorite except in the vicinity of the Rhobell Volcanic Group where stilpnomelane and chlorite predominate.

Interbedded siltstone and sandstone

In thin section the sandstones are moderately well sorted, fine to medium grained with rare, very coarse sand-sized angular grains. Most contain a high proportion of volcanogenic material. They differ from the overlying volcanic sandstone in being finer grained and containing relatively more matrix. Quartz makes up less than 10 per cent and feldspar, mainly altered, makes up 30 to 50 per cent. Lithic fragments mainly of feldspar-phyric lava make up to about 30 per cent. Matrix, 20 to 30 per cent, consists of sericite with some chlorite and varying proportions of calcite, which replaces the phyllosilicates of the matrix as well as feldspar clasts. Leucoxene or hematite/goethite are commonly concentrated in laminae.

Rare beds of quartz wacke, consisting of 35 to 45 per cent matrix minerals and a low lithic fragment and feldspar content (less than 10 per cent), occur.

The siltstone, commonly thinly interlaminated, is of two main types. One contains medium to coarse silt-sized quartz grains (0.02 to 0.05 mm in diameter) in a matrix of sericite with quartz and chlorite speckled with opaque minerals. The second type contains rounded pellets of chlorite, some rimmed by, or interleaved with muscovite. Grains of quartz and muscovite occur sparingly. The matrix consists of opaque minerals with sericite and chlorite (E 48515).

Volcanic sandstone

The thickly bedded coarse-grained sandstones are made up of reworked pyroclastic and epiclastic material and include tuffaceous sandstone and sandy tuffite with lithic tuff appearing mainly to the south of Cae'r-defaid.

In composition the tuffaceous sandstone varies from a feldspar-rich variety to a lithic rich variety grading into lithic tuffite and lithic tuff. Grains are subrounded to angular and sorting is poor to moderate. Quartz grains make up no more than 10 per cent and usually less than 5 per cent of these rocks. They are clear and probably of volcanic origin. They commonly show an overgrowth of authigenic quartz giving an euhedral outline (Plate 8.1). Feldspar clasts make up to 50 per cent and are up to about 0.3 mm long. Composition is generally albite but oligoclase and andesine also occur and in many beds the feldspar is highly altered. Lithic fragments, up to 40 per cent, are mainly of altered volcanic rocks, some showing a remnant trachytic texture in the groundmass, phenocrysts of feldspar or, more rarely, chloritic pseudomorphs after amphibole. Other lithic fragments include siltstone, rare quartzite, acid igneous rocks with perlitic fractures. Tricuspate shards, replaced by a quartz mosaic, are rare (E 42484). Opaque minerals are not generally abundant but locally numerous rounded detrital fragments of leucoxene (?after ilmenite) occur and may form distinct laminae. Magnetite occurs as rounded detrital grains showing both skeletal form and as euhedral crystals (E 45321). Pyrite is not common but occurs as discrete grains forming laminae or as irregular and euhedral overgrowths. Matrix, up to 25 per cent, is of variable amounts of chlorite, quartz and sericite after degraded lithic material. More rarely the matrix is of quartz or calcite.

Tuffite has been recorded most commonly, but not mapped as separate units, in the south-east from Coedwig to Cae'r-defaid. It has also been recorded low within the formation in Cwmhesgen, where a bed 1.10 m thick occurs some 40 to 45 m above the base of the formation. The tuffite ranges in grade from sandy to conglomeratic, and bedding features are similar to the tuffaceous sandstone. The tuffite mainly comprises fragments of feldspar-phyric lava similar to those of the lithic tuff (see below) but in addition contains a small proportion of other clasts. These include rhyolite, containing quartz phenocrysts; quartzite; clear, rounded euhedral quartz (E 43553), and fragments containing abundant iron oxides. The matrix consists of chlorite, sericite, calcite, with some minor epidote and opaque minerals. Some of the beds towards the top of the formation have a calcareous cement (E 44040).

The lithic tuff occurs at or near the top of the formation and is closely associated with the lithic tuffite of Coedwig/Cae'r-defaid which has probably been reworked from similar tephra. The rock displays crudely defined stratification and may show some grading. Grain sizes range from lapilli agglomerate to lapilli tuff, but bands of fine ash tuff also occur. Sorting is poor and clasts are angular to subrounded and generally in contact. Clasts of feldspar-phyric lava make up more than 75 per cent of the rock. These contain large

Plate 8
Photomicrographs of volcanic and sedimentary rocks in the Aran Volcanic Group.

1 Subgreywacke from Allt Lŵyd Formation composed of lithic clasts, altered feldspar and quartz with overgrowths. (E 47477, crossed polarisers, ×35).

2 Crystal tuff from Benglog Volcanic Formation composed of feldspar crystals, pumice fragments and chlorite. (E 42170, crossed polarisers, ×35).

3 Bedded hyaloclastite from Benglog Volcanic Formation showing scoriae with whole and squashed vesicles and rare feldspar crystals. (E 43919, plane polarised light, ×23).

4 Tuffite in which crystals of feldspar (upper middle), clasts of silicified rhyolite and devitrified shards are set in a muddy matrix. (E 42171, plane polarised light, ×35).

5 Pumice fragments in acid tuff from the Aran Fawddwy Formation. (E 43439, plane polarised light, ×30).

6 Devitrified glassy fragments in a hyaloclastite within the Benglog Volcanic Formation. (E 51691 plane polarised light, ×25).

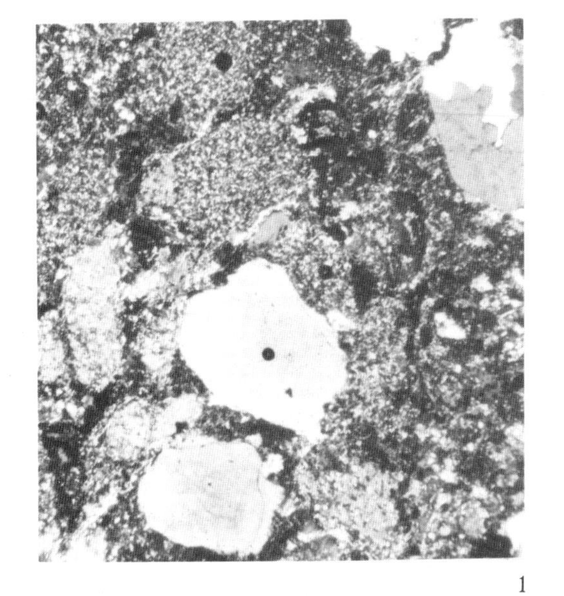
1

2

3

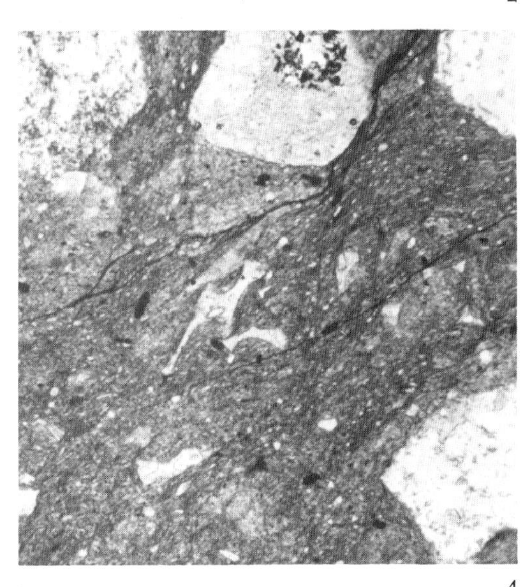
4

5

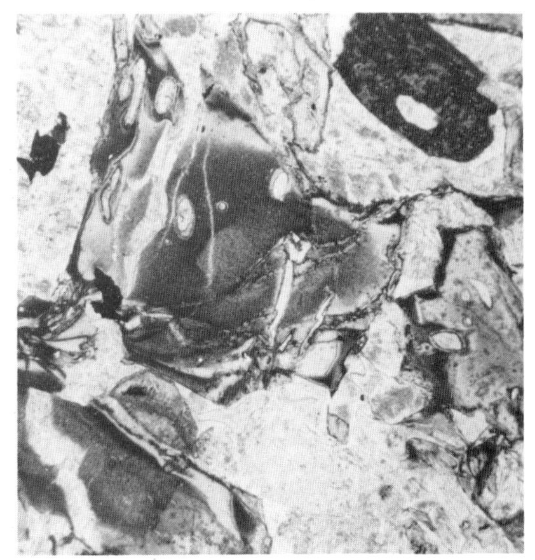
6

euhedral to subhedral phenocrysts of oligoclase and andesine which are zoned in places. Pseudomorphs of chlorite after ferromagnesian minerals occur more rarely. The groundmass of the porphyry consists of feldspar laths, and varying proportions of chlorite, quartz, euhedral sphene and disseminated leucoxene. Feldspar crystals, similar in shape and composition to those contained in the clasts, make up to 10 to 15 per cent. Shards are rare (E 44071). The matrix consists mainly of chlorite with some fine-grained feldspar. AAJ

Conglomerate (Aran Boulder Bed)

Conglomerate makes up the bulk of this member but breccia occurs in the lower part of the member and consists of poorly sorted angular to subrounded fragments of intermediate feldspar-phyric igneous material. On passing upwards the clasts become coarser and rounded, and the breccia grades into coarse massive conglomerate. The conglomerate is composed of subangular to well rounded pebbles, cobbles and boulders. The lower breccias and conglomerates are composed of matrix-supported clasts; on passing upwards these grade into better sorted more mature clast-supported conglomerates. PND

In composition the Aran Boulder Bed is mono-lithologic consisting of fragments of feldspar-phyric lava similar to those in the lithic tuffs and tuffite elsewhere within the Allt Lŵyd Formation. The lava fragments are intermediate in composition consisting of phenocrysts of feldspar with some mafic pseudomorphs after ferromagnesian minerals in a finer groundmass. The matrix comprises sericite with patches of recrystallised quartz, minor chlorite, scattered iron oxide and secondary epidote.

Minor lithologies

1 Volcanic rocks About 65 to 70 m above the base of the formation in Cwmhesgen, a thin bed of lapilli tuff consists of angular basaltic fragments in a calcareous cement. On Foel Offrwm, a bed of lithic tuff, consisting entirely of fragments of acid tuff or rhyolite,occurs at the top of the formation [7500 2132] (E 47515).

2 Oolitic mudstone Thin beds of oolitic mudstone (Plate 9.1) occur in places but are difficult to trace laterally. One of these is exposed in Cwmhesgen about 90 m above the base of the formation and a similar bed is present at an equivalent stratigraphic level near Twr-y-maen, about 1 km to the north, where it is associated with graptolitic mudstone. In Cwmhesgen the succession from top to bottom is as follows:

	Approx. thickness m
Volcanic sandstone, grey cross-bedded with streaks of dark siltstone	-
Lapilli tuff, angular basaltic lapilli in calcareous groundmass. Thick vertical calcite veins in complete exposure	0.5
Black siltstone	1 to 2
Oolitic mudstone	4
Exposure gap with scattered siltstone and sandstone	17
Sandy tuffite	1.10
Interbedded siltstone and sandstone	17
Dolerite (intrusion)	1
Interbedded siltstone and sandstone	25

In the oolitic bed near Twr-y-maen the ooliths range from approximately 0.15 to 4 mm in diameter. Most are simple and ovoid, but some are compound comprising two or more small ooliths. They consist of concentric layers of phyllosilicate accompanied by variable proportions of iron oxides. Generally the ooliths lack obvious cores but some contain central patches of polycrystalline quartz, some of which are intergrown with opaque minerals. Some ooliths show signs of secondary recrystallisation in the form of porphyroblasts overprinting the primary growth rings. The matrix consists of a variable ratio of iron oxide and green phyllosilicate permeated by orientated anastomosing laminae of opaque black material which is possibly calcareous. X-ray diffraction analysis by K. S. Siddiqui shows that the ooliths consist of chlorite, goethite and quartz (NEXD 1181), and the matrix is of a similar composition (NEXD 1182) with no representation of the composition of the opaque matter.

Elsewhere the ooliths contain a central nucleus of altered feldspar and the mudstone contains scattered crystals of feldspar and clear quartz of volcanic origin (E 43558).

On Foel Offrwm a lens of oolitic mudstone (maximum thickness 50 cm) occurs at the top of the formation resting on volcanic sandstone [7475 2102], and north of Cae'r-defaid [7982 2389] a siltstone at this horizon contains numerous algal concretions. These are generally compound structures 1 to 2 cm in diameter and consist of alternating laminae of chlorite and collophane (E 50439). They are thought to be indicative of shallow water conditions (R. W. O'B. Knox, personal communication).

3 Breccias A horizon of slumped breccia is developed locally at the top of the Allt Lŵyd Formation on the west slopes of Dduallt

Plate 9 Photomicrographs of volcanic and sedimentary rocks in the Aran Volcanic Group.
1 Ooliths of chlorite, goethite and quartz in a tuffaceous silty matrix containing disseminated carbonaceous material.
2 Dolerite intruded into base of pillow lava, showing strained clinopyroxene, plagioclase and leucoxenised ilmenite. (E 44066, crossed polarisers, ×24).

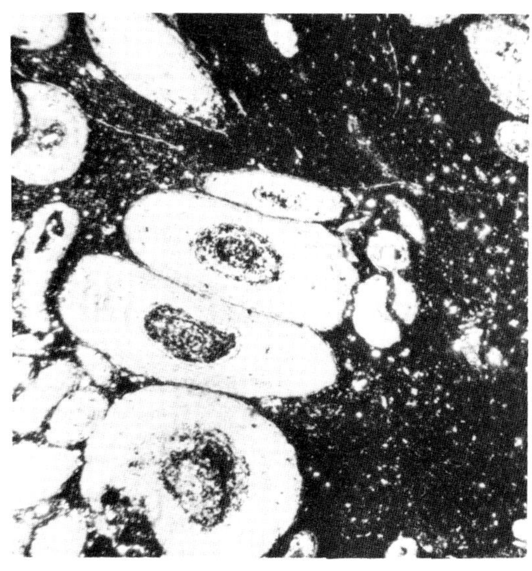

1

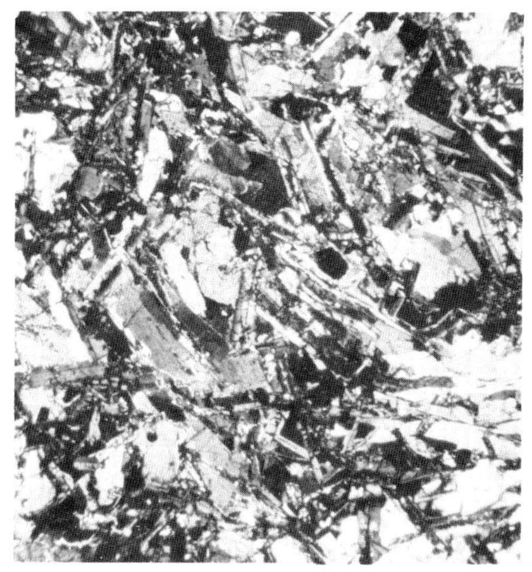

2

[8051 2668] and on Ffridd yr Allt-lŵyd. The breccia contains blocks of volcanic sandstone and quartz-rich subgreywacke, together with banded siltstone and mudstone, in a muddy matrix. On Ffridd yr Allt-lŵyd [7978 2980], north of the type section, where the breccia is best developed, it is about 23 m thick and rests on volcanic sandstone. It passes laterally (but is separated by a fault) into 98 m of contorted slumped siltstone, which contains scattered lithic fragments and beds of crystal rich siltstone. The breccia and slumped siltstone mark the southern limit of the condensed sequence where the Benglog Volcanic Formation rests directly on the Allt Lŵyd Formation; the thickest part of the deposit is bounded by downthrow faults, which may have formed a narrow rift controlling the deposition of this mudflow or gravity flow deposit at the edge of the basin.

Underlying the banded siltstone of the Allt Lŵyd Formation near Blaenau [792 231] is a lens shaped body of unsorted pebbly mudstone (E 49337, 49338) which consists of angular fragments of siltstone and quartzose siltstone, similar to the lithologies seen in the underlying Ffestiniog Flags and Cwmhesgen formations. The fragments are held together in a silty matrix. The sediment is probably the result of mass flow of contemporaneously eroded Cambrian sedimentary rocks.

Strata between the Allt Lŵyd Formation and the Offrwm Volcanic Formation

Locally the Allt Lŵyd Formation is overlain by a thin siltstone which is generally poorly exposed. It is dark grey and generally tuffaceous containing scattered rounded euhedral and fragmented quartz and feldspar crystals (Plate

Upper Flow Unit on Foel Offrwm

Upper part

Middle part

Lower part

Lower Flow Unit on Foel Offrwm

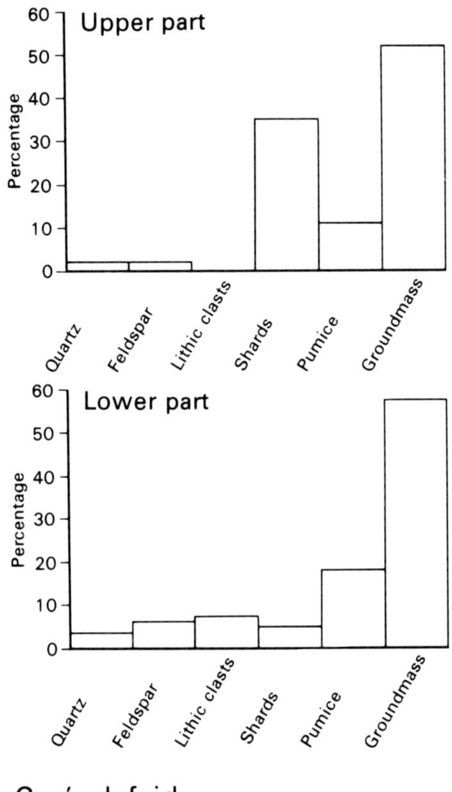

Cae'r-defaid

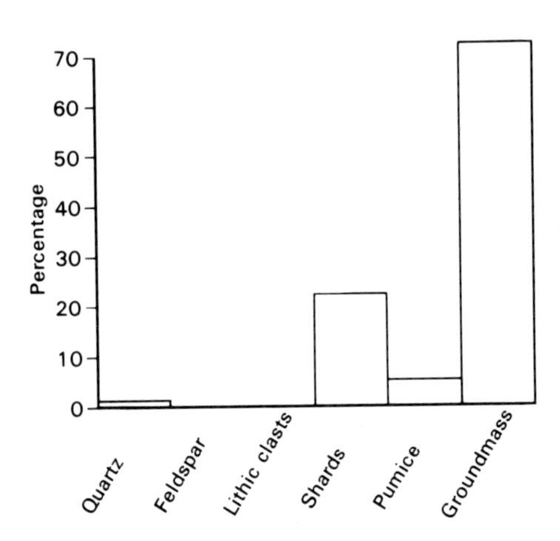

Figure 10 Histograms, based on point count analysis of thin sections showing the vertical and lateral components within ash-flow tuffs in Offrwm Volcanic Formation

8.4) and small altered angular lithic clasts. Some lithic clasts show a devitrified shardic texture and others contain feldspar phenocrysts. These volcanic clasts are contained in a ground mass of chlorite and sericite, which in places is rich in collophane (E 50439).

OFFRWM VOLCANIC FORMATION

Within the district the Offrwm Volcanic Formation is thickest, about 190 m, on the north-west slopes of Y Foel [770 180], where it includes rhyolitic lavas. On Moel Cae'r-defaid it is represented by 7 to 8 m (Figure 7) of fine-grained tuff which wedges out to the north. The formation correlates with the Lower Acid Volcanic Group, which is best developed on Mynydd-y-Gader (Cox and Wells, 1927) to the south of the district.

The formation consists mainly of acid tuff and represents the first major acidic eruption in the district. Two tuffs can be distinguished. The lower, probably a pyroclastic flow deposit, rests on shallow water sediments and shows some welding. The upper tuff contains tubular pumice and shows little evidence of welding. It is massive at the base and thinly bedded near the top with crystals and pumice reworked into alternating thin bands. It may have been emplaced by a pyroclastic flow or as a secondary debris flow of unconsolidated ash triggered by earth movements.

DETAILS

Type section

In the type section on Foel Offrwm the formation is about 91 m thick and comprises two tuff units. There are no rhyolite lavas.

The lower tuff thickens from 22 to 30 m from west to east. It contains more crystal and lithic clasts than the upper tuff. Both tuffs are massive at the base, and thinly bedded, vitric and crystal lithic at the top. Concentrations of pumice fragments accentuate the bedding. In the upper tuff, a foliation parallel to bedding becomes more common near the top. The composition of the tuffs (Figure 10) shows an upward increase in the shard content in both flows.

Lithologies

The tuff on Foel Offrwm is typically greyish olive green weathering to a yellowish grey. Crystals of quartz and feldspar, up to about 1 mm in length, are euhedral, or fragmented and resorbed. Crystals of clear, deeply embayed quartz are considerably more abundant than feldspar. The feldspar (oligoclase) is normally twinned and cloudy containing numerous dusty inclusions of ?sphene. Mafic pseudomorphs are rare. Lithic clasts occur mainly near the base of the lower tuff and rarely in the upper tuff. Most are rounded, up to 2.5 mm long, and consist of cryptocrystalline chlorite, sericite and quartz with flecks of iron oxides which are commonly concentrated in the margins of the clasts. One fragment in the upper tuff [7508 2131] is over 60 cm long and consists of phenocrysts of quartz, surrounded by recrystallised vitric material, and rarer feldspars with mafic pseudomorphs after ?amphibole in a recrystallised groundmass of sericite and chlorite showing a relict shardic texture (E 48425). Tubular pumice fragments with feathered terminations are infilled with sericite, quartz and chlorite. In the lower tuff the pumice clasts are slightly moulded around clasts and crystals and the tuff is welded locally. In the upper tuff pumice is relatively more abundant and there is no evidence of welding. In both tuffs cuspate, tricuspate and bubble forms of shards replaced by a mosaic of sericite, quartz and rare feldspar are abundant. The groundmass consists of sericite, quartz and chlorite, with scattered blebs of opaque minerals. Irregular patches of calcite occur within the lower tuff.

North-east of Cae'r-defaid the tuff is thinly banded and consists of shards with small fragments of pumice and rare quartz phenoclasts (E 43916) in a vitric groundmass. It is interpreted as the distal part of either one or both of the Foel Offrwm tuffs. VMJ

On Y Foel the rhyolites are blue-grey, weathering white, and are flow-banded, flow-folded and brecciated. The breccias are autoclastic comprising subangular rhyolite blocks supported in flow-banded rhyolite, and are interpreted as flow-breccias extruded as a dome. The rhyolites contain microphenoclasts of quartz and albite in a finely crystallised groundmass of quartz, feldspar and sericite exhibiting a trachytic fabric and, in places, perlitic cracks. Above the rhyolites on Y Foel and to the north-east, the ash-flow tuffs are welded and contain crystals (up to 2 mm) of albite and abundant quartz. The groundmass is recrystallised but some shards are discernible. PND

MELAU FORMATION

In the type locality in Afon Melau the formation is an estimated 220 m thick and rests on the Allt Lŵyd Formation. At Cae'r-defaid, less than 2 km to the north-west it is 33 m thick and overlies the Offrwm Volcanic Formation. A further 1.5 km in this direction the formation wedges out. South-westwards from Afon Melau the formation is exposed in the core of the syncline at Buchesydd [7798 2107] and on Foel Offrwm (Figure 11). The formation does not appear to have an equivalent in the Cader Idris area, but 3 to 4 m of basic tuff [7941 2047] (not shown on the map) near Pont Sêl, south of the Wnion, may be its lateral equivalent.

The formation consists of basic tuff, tuffite and thin lavas (Figure 11). The tuff and tuffite weather deeply and are poorly exposed. Cox and Wells (1927) suggested that the basic rocks on Foel Offrwm were possibly equivalent of their more extensive Lower Basic Group on Cader Idris, which overlies siltstone yielding *Didymograptus 'bifidus'* of Llanvirn age. There is no indication of the age of the beds from Foel Offrwm, but on Moel Cae'r-defaid the basic tuff underlies the Llanvirn siltstone, and south-west of Afon Melau the record of *Merlinia selwynii* (p.32) indicates an Arenig age. However, details of this latter locality are vague and it is possible that the fossils were derived from a block within the tuffs.

The Melau Formation represents the first episode of basic volcanicity in the Aran Volcanic Group. The formation includes mudflow deposits and tuffite, which are more widespread than the basic tuff. These deposits are interpreted as the product of mass movement of unstable accumulations of basic tuff, possibly stimulated by seismic shock. From the variation in thickness and bedding the eruptive centre was probably close to Afon Melau.

DETAILS

Type section

In Afon Melau [7961 2229] the formation comprises basic tuffs with a few thin intercalated basic lavas. In adjacent ground less than

1 km to the north, the tuff rests on an estimated 150 m of basic tuffite consisting of rounded fragments of vesicular basalt and scattered feldspar crystals in a muddy matrix (E 44073). This tuffite and the thin siltstone that overlies it are absent in the Melau gorge where the basic tuff rests directly on the Allt Lŵyd Formation.

At the base is a coarse lapilli tuff with interbedded agglomerate. The lapilli tuff is moderately well bedded, 30 to 80 cm thick, with alternating coarse and fine bands, some of which are graded. Poorly developed low angle cross-stratification indicates reworking. In general the tuff fines upwards and becomes more massive, although agglomeratic lenses persist. A disused quarry on the west bank of the stream [7961 2192] exposes 9 to 12 m of tuff with only one bedding plane discernible.

Lithologies

1 Afon Melau The basic tuff mainly consists of angular fragments of vesicular basalt which suggests fracturing of a solidified magma (E 47534). Elsewhere the fragments have chilled margins (E 48200), or are defined by a layer of stretched vesicles (E 47534) indicating incorporation in a molten state. The basaltic clasts comprise feldspar laths within an aggregate of chlorite and leucoxene. Vesicles are filled with chlorite, calcite or quartz and epidote (E 48200). The matrix is mainly of calcite but with sericite and quartz mosaic in places. Leucoxene/sphene is common, but pyrite and magnetite are rare.

Glassy basaltic lavas interbedded with the tuffs are generally much altered. Most consist of albitised feldspar laths, 0.3 to 0.4 mm long, with interstitial chlorite, abundant leucoxene/sphene and patches of secondary quartz and calcite. Stilpnomelane occurs in places (E 43903). Vesicles are infilled with chlorite and some have a rim of quartz.

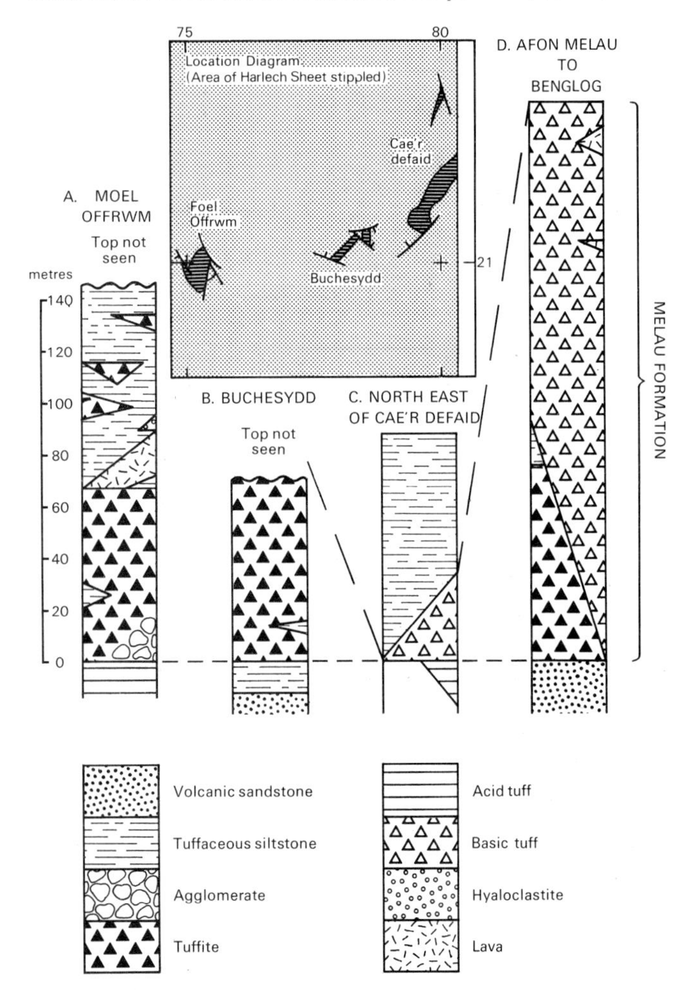

Figure 11 Generalised vertical sections showing lateral variations in the Melau Formation

2 Cae'r-defaid The basic tuff here is similar to that seen in the Afon Melau but contains clasts consisting of clusters of tabular feldspar crystals and lapilli (E 44045).

3 Foel Offrwm The Melau Formation is relatively well exposed and consists mainly of basic tuffite and lavas with intercalated tuffaceous siltstone (Figure 11). Locally, in the south-west, the lowest 18 m is a breccia, comprising angular blocks of flow-banded and massive rhyolite, up to 1 m long, in a lithic crystal matrix which is basic in composition and similar to that of the overlying basic tuffite. The breccia was probably emplaced as a mudflow derived by erosion of a rhyolitic dome to the south.

The overlying rocks consist of interbedded breccia and conglomeratic tuffite with intercalated conglomeratic sandy tuffite. In general, bedding is poor in the lower part, but bands and laminae of fine-grained tuffite occur towards the top. The tuffite consists predominantly of fragments of porphyritic basic lava and scoria with feldspar clasts. In places fragments of rhyolite, acid tuff and siltstone are abundant. The matrix consists of a fine-grained mosaic of quartz, sericite, leucoxene/sphene and possibly feldspar (E 48437).

The finer-grained tuffite in the upper part of the formation contains a greater variety of lithic material, and the crystal content varies from about 20 to 60 per cent. Crystals are predominantly of altered oligoclase up to 1.75 mm long, and quartz crystals of volcanic origin are rare (less than 1 per cent). Scoria consisting of altered feldspar phenocrysts in a chloritic matrix are commonly lens-shaped in outline with feathered terminations. Other clasts consist of large tabular feldspar phenocrysts held together by a chloritic matrix, feldspar-phyric lava, basalt, recrystallised pumice, rhyolite and siltstone. Ooliths and devitrified shards are a minor component. The matrix is crypto-crystalline with much leucoxene.

There is considerable lateral as well as vertical variation within these tuffites. On Foel Offrwm the lower rhyolitic breccia dies out northwards in less than 1 km, though small fragments of perlitic rhyolite continue at this level and are locally abundant (E 47513). On the north side of the hill chlorite scoria are common and locally make up most of the clastic material. These are fragmented, and some show concentric bands of sericite or less commonly of calcite. In addition to feldspar phenocrysts some contain flattened vesicles. Zircon is a common accessory mineral (E 45333).

To the north-east of Foel Offrwm at Buchesydd the tuffite is interbedded with impersistent tuffaceous siltstone. Crystals include broken and rounded euhedral augite singly or in clusters, and edged with tremolite/actinolite (E 45328), quartz (less than 5 per cent), and feldspar. Unaltered feldspar crystals range from andesine to labradorite but most are extensively altered and replaced by sericite. Most lithic clasts have been derived from basaltic lavas or hyaloclastite, and contain feldspar phenocrysts and rarely euhedral pyroxene in a chloritic matrix. Vesicles are infilled with chlorite or a quartz mosaic. Fragments of acid lava, tuff and feldspar-phyric lava occur, in addition to clasts consisting predominantly of iron oxides. Anhedral pyrite occurs sparingly.

On the south side of Foel Offrwm, 24 m of basalt overlie the tuffite. The basalt is massive in the lower part, vesicular, locally pillowed [7501 2070] in the upper part, and with some hyaloclastite at the top [7515 2072]. AAJ

Strata between the Allt Lŵyd Formation and the Brithion Formation

South of Afon Wnion, siltstone with intercalated volcanic rocks, 100 to 150 m in thickness, overstep the Offrwm

Volcanic Formation to rest on the Allt Lŵyd Formation to the east of the district. The rocks become pisolitic and eventually wedge out.

The rocks are exposed in Afon Celynog downstream from Pont Sêl [7950 2045]. The sequence begins with 3 to 4 m of brown weathering basic tuff which may be the lateral equivalent of the Melau Formation. This is overlain by 15 m of dark pyritous mudstone and 25 to 30 m of poorly sorted polymict breccia with thin mudstone intercalations. The breccia contains angular to subangular fragments of basic lava, rhyolitic volcanic material, chert, mudstone, quartz grains and numerous feldspar crystals, supported in a structureless matrix of tuffaceous mudstone. The volcanic clasts are up to 10 cm in size, although most are in the order of 1 to 3 cm. The mudstone clasts are mostly lenticular reaching up to 30 cm in length and commonly lie parallel to bedding. The breccia is overlain by 2 to 3 m of grey-brown, cross-bedded tuffaceous sandstone, which passes up into approximately 70 m of finely bedded (strongly cleaved) black pyritous micaceous mudstone with thin intercalations of fine tuffaceous sandstone and mud-rich polymict breccia similar to those described above. Downstream from Pont Sêl the mudstone is locally very fossiliferous and has yielded graptolites which suggest the Llanvirn *D. bifidus* to ?*D. murchisoni* zones.

At the top of the formation the mudstone is overlain by a few metres of fine sandstone followed by about 10 m of poorly sorted polymict breccia containing a very variable assortment of acid and basic volcanic clasts, sandstone fragments and mudstone clasts, supported in a mudstone matrix. The breccias are interpreted as the product of mass movement of unstable epiclastic volcanic material and mud by debris flow. Features supporting this interpretation are the poorly sorted and structureless nature of the deposits, the matrix supported habit of the clasts, and their heterolithic nature.

BRITHION FORMATION

The formation is confined to the south-eastern part of the district where it crops out on both sides of the Bala Fault. It consists of up to 75 m of volcanic breccias and acid ash-flow tuff, and is the lateral equivalent of the Creigiau Brithion Formation (Dunkley, 1978) in the Aran Mountains where it is interpreted as the remains of an extrusive rhyolite dome. It wedges out to the south-west and to the north.

DETAILS

The tuff shows a sharp, conformable contact with the underlying mudstone. The basal breccia contains angular clasts of non-vesicular rhyolite and mudstone. The latter, up to 25 cm long [7848 1893], are lenticular to spherical in shape with rounded edges. They were probably derived from the underlying sediments during flow. In one section [8030 1895] the basal breccia is overlain by approximately 20 m of welded acid ash-flow tuff containing flattened pumice lapilli and exhibiting a marked eutaxitic fabric. The tuff is welded throughout, and at the top the eutaxitic fabric is truncated by a discordant erosional surface. This erosional surface is well developed on the south-east side of the Bala Fault and is present at several localities to the east of the district within the Arans (Dunkley, 1978).

Strata between the Brithion Formation and the Benglog Volcanic Formation

South-east of the Bala Fault on the southern edge of the district [7915 1785] the Brithion Formation is locally overlain by siltstone with associated basic tuff, breccias and tuffaceous sandstone with some thin basic lavas near the top. The breccias are composed of basic lava clasts with a few clasts of rhyolite, in a calcareous tuffaceous groundmass. The sandstone is coarse-grained, cross-stratified, ferruginous and highly calcareous; it weathers deeply on exposure. PND

North of the Bala Fault, where the siltstone oversteps both the Brithion and Melau formations to rest on the Allt Lŵyd Formation, it is locally up to 170 m in thickness and thins northwards to about 5 m just north of Cwm yr Allt-lŵyd (Figure 7). Northwards the siltstone passes laterally into slumped deposits which wedge out to the north, so that in the northern part of the district the Benglog Volcanic Formation rests directly on the Allt Lŵyd Formation.

BENGLOG VOLCANIC FORMATION

This, like the Allt Lŵyd Formation, extends from north to south across the whole district. It is the Basic Volcanic Group of Wells (1925). The formation is about 240 m on the southern margin of the district, and about 230 m thick in the north near Guog Duon, but shows considerable variation in thickness between these two localities.

Crystal tuff makes up the greater part of the formation and, as with the Melau Formation, much of the tuff may be secondary pyroclastic debris emplaced by sediment gravity flow. The tuff contains intercalated lavas and siltstone. The crystal tuff is thickly bedded with a poorly defined stratification discernible in places, and consists predominantly of altered and fractured phenoclasts of feldspar in a dominantly chloritic groundmass. Xenoclasts of black siltstone are common, especially in the northern part of the district, and clasts of porphyritic microtonalite (quartz-latite of Lynas, 1973) are rarer but widely scattered. South of the Afon Wnion some lava flows preceded the main tuffs, but elsewhere a substantial volume of tuff was deposited before the eruption of any lava. From the thickening of the lavas and their association with hyaloclastites the locations of at least two eruptive centres can be suggested. One, on the western slopes of Craig y Benglog, shows a pillow lava (Plate 7.1) intruded by a thick dolerite sill and overlain by a wedge of bedded hyaloclastite. A thin NW-trending dyke in the tuffs below the sill may have acted as a feeder. The second centre, on Ffridd yr Allt-lŵyd, was active somewhat later than the first, and on Allt Lŵyd a lava overlies 40 to 60 m of hyaloclastite material. Both units thin away from the area, although the hyaloclastite wedges out more abruptly than the lava.

The broken euhedral feldspars of the tuff contrasts strongly with the lath-shaped feldspars of the lavas and it is clear that tuff and lavas originated from different parent magmas.

DETAILS

Type Section

On the western slopes of Craig y Benglog the formation consists of coarse crystal tuff with scattered blocks of siltstone and intrusive igneous rocks overlain by pillow lavas which in turn are overlain by bedded crystal tuff of more variable composition. At its base [8013 2408] the formation rests on tuffaceous siltstone which shows a marked increase in the crystal content in the highest few metres; locally, the tuff shows basal load-induced flame structures.

On Craig y Benglog the lower crystal tuff varies from about 50 to 80 m within less than 1 km. The tuff is generally massive and coarse grained, becoming increasingly well bedded with low angle cross-stratification in the uppermost 30 m and rare intercalated fine-grained tuff.

The basaltic pillow lava is reddish brown and can be traced for 2 km. Locally, on Benglog two pillow lavas, each about 13 m thick, are separated by a dolerite sill [8025 2410]. The upper pillow lava passes laterally into a wedge of bedded hyaloclastite which also cuts out the dolerite. The contrast between the upper pillow lava and the dolerite is transitional and rubbly, suggesting that they were contemporaneous. To the north the pillow lava is 15 to 20 m thick and contains pods of ophitic dolerite mainly at the base.

The upper tuffs of the formation are bedded and reworked. Minor channels, less that a metre across, occur rarely. One channel [8050 2467] is filled with coarse feldspar crystals, with interstitial finer-grained feldspar, chlorite, quartz and calcite forming less than 10 per cent of the rock. Composition of the tuffs is variable and some are more acidic, containing acid shards and rare quartz phenoclasts. There is an increase in the ratio of scoria to crystal clasts towards the top, and intercalated accretionary lapilli tuff appear locally. Ridgway (1971) commented on a similar increase in acid shard content towards the top of the crystal tuff in the Braich-y-ceunant area [797 202] and concluded that eruption took place from a differentiating magma chamber. Thin basaltic lavas are impersistent but fairly common, and some contain fragments of chilled vesicular lava. On the south side of Craig y Benglog the upper part of the sequence is intruded by a thick dolerite sill, but to the north, on Geological 1:50 000 Sheet 136, the more acidic nature of the top of the formation is again apparent. Locally a coarse volcanic breccia, up to 5.5 m thick, consists of abundant blocks (to 40 cm long) of tuffaceous siltstone in coarse crystal tuff. AAJ

Local variations

South-east of the Bala Fault the formation is poorly exposed. At the top there is a thin welded ash-flow tuff of andesitic-dacite composition.

North of the Bala Fault, around Ty-newydd-uchaf [7885 1940], the formation is about 150 m thick and contains a basic lava at the base. The overlying sequence comprises crystal tuff and thin mudstone, with basic and andesitic lavas [7883 1911] at the top. Eastwards in Afon Celynog around Braich-y-ceunant [7963 2024] the formation contains few lavas. The upward succession comprises 100 m of crystal tuff, 30 m of thinly intercalated mudstone, basic lavas, hyaloclastite, crystal tuff and poorly sorted basic volcanic breccias, and 70 m of massive hyaloclastite, tuff and polymict volcanic breccia with mudstone intercalations. PND

Northwards from Craig y Benglog the thin lavas mostly wedge out but the pillow lava persists for about 2 km on Geological 1:50 000 Sheet 136. To the north the formation consists of fairly uniform crystal tuff, 110 to 120 m thick, with little intercalated siltstone. North of the Mawddach on Ffridd yr Allt-lŵyd the formation rests on a thick sequence of slumped siltstone and tuffaceous siltstone. Here the sequence is more variable and the crystal tuff is thin (Figure 7). The upward succession comprises 28 m of crystal tuff, 30 m of mudflow breccia, up to 70 m of hyaloclastite overlain by 40 to 70 m of pillow lava. The hyaloclastite can be traced for about 1½ km and is overstepped to the north and south by the basalt which has been traced for a total of 8 km near the eastern limits of the district forming the local top to the Benglog Volcanic Formation.

Northwards from Ffridd yr Allt-lŵyd to the edge of the district the crystal tuff again increases in thickness from 90 to 140 m and includes two 10 m bands of tuffaceous siltstone, one about 30 m from the base and the second at the top of the crystal tuff and overlain by the basalt. The tuffaceous siltstone units can be traced for 2 and 1 km respectively and contain numerous shards, feldspar phenoclasts and some mafic pseudomorphs in a silty matrix.

Lithologies

1 Crystal tuff This constitutes the largest part of the formation. The tuff (Plate 8.2) is fairly uniform, dark greenish grey and andesitic in composition. It consists of feldspar crystals, scattered lithic clasts and bombs in a predominantly chloritic matrix. Feldspar crystals, comprising 80 per cent of the tuffs, are rounded euhedral and show an intricate pattern of fractures. Composition is in the oligoclase to andesine range but most show varying degrees of alteration to sericite or replacement by calcite. Most of the crystals show no twinning, although some simple Carlsbad, albite and combined Carlsbad/albite and pericline twinning occur.

Lithic clasts are a minor but conspicuous component. Rounded black siltstone fragments are the most common but rarely exceed 2 to 3 per cent. Large rounded blocks, some 30 × 60 cm, of microtonalite are rare but widely scattered. They consist of scattered euhedral albite phenocrysts in a groundmass of fine acicular feldspar and interstitial quartz, with vesicles filled with chlorite or prehnite (E 43772); the rock compares closely with that of the intrusion on Ffridd yr Allt-lŵyd [8010 2900], and is similar to the quartz-latite described by authors working to the north of the district. Large chloritic fiamme and, more rarely, clasts of basalt have also been recorded. In the lower part of the formation some clasts consist of tabular feldspars in a chloritic groundmass.

Matrix consists mostly of chlorite and calcite but more rarely of a mosaic of quartz, chlorite and calcite. Some stilpnomelane occurs together with dusty leucoxene and sphene.

In the upper part of the sequence the crystal content of the tuff decreases with a proportional increase in scoria and fine-grained matrix. In these beds shards can be distinguished in places and some beds are rich in scoria fragments. These are irregular in shape, and fractured vesicles along the margins suggest brittle fracture. The internal texture is accentuated by leucoxene and the vesicles are infilled with chlorite, calcite or more rarely with quartz. Some contain a few feldspar phenocrysts. Interbedded reworked tuffs show a relative winnowing of the fine-grained material and a concentration of crystal clasts or, in places, scoria. These are sorted into laminae of medium and coarse grade material (E 48453) with some laminae showing concentrations of magnetite, pyrite and ?chalcopyrite. This reworked material may occur in separate beds or as the upper part of an otherwise massive tuff. Clear, rounded quartz clasts of volcanic origin form a minor (less than 2 per cent) but conspicuous component and, in addition, some small angular quartz grains (less than 5 per cent) may have been derived from a 'sedimentary' source (E47537). Lithic fragments (less than 10 per cent) are a recrystallised mosaic of quartz and chlorite with some showing perlitic fractures.

2 Tuffaceous siltstone and mudstone Thin persistent beds of tuffaceous siltstone and mudstone occur within the reworked tuffs and sparsely within the crystal tuff. These beds consist of varying proportions of scattered large euhedral and generally altered feldspars, pseudomorphs of chlorite and sericite possibly after feldspar, chloritic fiamme, devitrified cuspate and tricuspate shards, and a variety of lithic clasts in a siltstone or mudstone matrix.

3 Acid tuff Thin beds of acid tuff are rare and are confined to the upper part of the formation. They generally consist of a recrystallised mass of quartz and chlorite but some contain scattered feldspar. Shards and tabular pumice are discernible (E 43436) in places.

4 Lavas The two thickest and most extensive lavas mapped within the district are centred on Benglog and Ffridd yr Allt-lŵyd; both have well developed pillows. Thinner lavas are massive, columnar jointed and blocky. The lavas are fine to medium-grained and variolitic: in the thick flows they are coarser grained and subophitic and are texturally indistinguishable from dolerite. The distribution of vesicles is variable although in the massive flows they are commonly concentrated towards the top. They are usually filled with combinations of chlorite, calcite, epidote, clinozoisite, quartz and prehnite.

On Craig y Benglog the pillow lava is comparatively fresh. It consists of slender lath-shaped crystals of calcic andesine (c. 50 per cent), commonly altered to sericite, with interstitial pale green augite (c. 37 per cent) and chlorite. Augite is anhedral to subhedral and in places has a sub-ophitic texture with the feldspar (E 44068). Large pseudomorphs of chlorite with sericite and minor quartz after ferromagnesians occur sparingly. Secondary carbonate and quartz is patchy and the lava is densely speckled with sphene/leucoxene. The associated dolerite (Plate 9.2) is coarser and shows well defined ophitic texture with fresh feldspar in the calcic andesine to labradorite range (E 44055) but are otherwise similar to the lava.

In the lava on Ffridd yr Allt-lŵyd the feldspars are albitised and the pyroxenes have been altered to chlorite and stilpnomelane. The thinner flows occurring in places generally contain altered feldspars and pyroxenes which have been replaced by tremolite/actinolite and chlorite. Secondary epidote and stilpnomelane are less common.

The blocky lavas contain blocks which are texturally distinct from the host lava either in crystal size or in the concentration of vesicles. The outline of the blocks is generally sharp but irregular, suggesting that they formed during flow.

The hyaloclastites (Plate 8.3, 8.6) are variable, but typically contain blocks of vesicular basalt in a recrystallised chloritic groundmass densely speckled with opaque minerals. AAJ

Strata between Craig y Ffynnon Formation and Benglog Volcanic Formations

South-east of the Bala Fault the Benglog Volcanic Formation is overlain by 60 m of uniformly grey siltstone and mudstone, which in turn are overlain by an impersistent fine-grained basaltic lava about 25 m thick.

CRAIG Y FFYNNON FORMATION

This rests partly on the siltstone and partly on the basalt. South of the Afon Wnion and north-west of the Bala Fault exposure above the Benglog Volcanic Formation is poor. Farther north, just to the east of the district, the Craig y Ffynnon Formation is overstepped by the Aran Fawddwy Formation on the southern slopes of Dduallt.

At Craig y Ffynnon [832 195], to the east of the district, the type section for the formation, described by Dunkley (1978), consists of 125 to 150 m of rhyolite overlain by welded acid ash-flow tuffs. At the top the tuffs are eroded and overlain by the Pistyllion Formation. Within the district the formation is poorly exposed and estimated at about 40 m thick. The rhyolite wedges out to the south-west so that only the ash-flow tuff is present. The tuff contains tubular pumice and is not welded at the base.

PISTYLLION FORMATION

The type section for this formation (Dunkley, 1978) lies near Pistyllion [8225 1900] and Bwlch y Fign [8820 1855] to the east of the district. Here it is approximately 300 m thick and consists mainly of basaltic and andesitic lavas with some mudstone, mudflow breccias and crystal tuff.

Within the Harlech district the formation is not exposed but an estimated 70 to 80 m occurs under drift in the south-east. PND

ARAN FAWDDWY FORMATION

The formation correlates with part of the Upper Acid Group of Cox and Wells (1927). It takes its name from the highest peak of the Aran Mountains where its full thickness is in the order of 350 to 400 m. Dunkley (1978) described the type section in this area. Within the Harlech district it occurs in the extreme south-east under drift, where it rests on the Pistyllion Formation, and northwards between the Afon Mawddach and Afon Lliw where it rests on basalts of the Benglog Volcanic Formation. The formation consists of acid ash-flow tuffs, but in the Aran Mountains rhyolite occurs at its base. North of Afon Mawddach the formation is relatively well exposed.

DETAILS

The tuff is generally massive but in places contains poorly defined bedding, the beds usually being over 1 m thick. It contains scattered albite and oligoclase in a recrystallised mosaic of quartz and feldspar with varying proportions of chlorite, sericite and opaque minerals. The feldspars are commonly pseudomorphed by sericite and quartz, and some are surrounded by a film of recrystallised glass. Quartz phenoclasts are rare. Lithic clasts are sparse although local concentrations occur near the base. Clasts are generally acidic but a few are of a more basic composition. Where recrystallisation has not obscured the original texture, shards and pumice fragments (Plate 8.5) are abundant. Within the district the tuff is not welded, but to the south, in the Aran Mountains, strongly welded textures have been recorded (Dunkley, 1979). The tuff is generally uniform in composition and appearance. Intercalated breccia and fine ash-grade tuff containing no crystal clasts are rare. To the east, on 1:50 000 Geological Sheet 136 (Bala), the tuff contains about three intercalations of siltstone and basaltic breccia. AAJ

ORDOVICIAN BIOSTRATIGRAPHY

Bolopora undosa (Lewis, 1926), recorded from the Garth Grit Member of the Allt Lŵyd Formation was originally said to be a bryozoan, but recent work by Hofmann (1975) has shown that it is probably not organic in origin (p. 33).

The interbedded siltstone and sandstone beds of the Allt Lŵyd Formation have yielded no identifiable fossils during the present survey. Just to the north-east of the Harlech district, however, the lateral equivalents of these beds (Fearnsides' Llyfnant Flags) have yielded graptolites of Arenig age: Fearnsides (1905, p. 619) referred to the presence of *Didymograptus deflexus* and *D. extensus*, and Zalasiewicz (1984) has revised these as *D.* aff. *deflexus* and *D.* aff. *simulans*. These indicate the Zone of *D. extensus*.

At higher levels the volcanic sandstone facies of the Allt Lŵyd Formation has yielded shelly fossils at a few localities from Moel Llyfnant to the valley of the Wnion. At four localities in the north-east corner of the district, between Nant-ddu and the south-west flank of Moel Llyfnant, beds of siltstone and tuffaceous sandstone have yielded the following: *Monobolina sp.*, *Orthambonites proava*, *Redonia*?, *Sinuites*?, orthocone [a possible Endoceroid allied to '*Conoceras*' *eoum*], *Merlinia selwynii* and *Nesuretus parvifrons*. These are typical of the 'Henllan Ash' of the Arenig area. From localities just east of the district, Whittington (1966) described a 15 m unit of 'Henllan Ash', yielding a trilobite fauna including *Ampyx cetsarum*, *Merlinia selwynii*, *Myttonia fearnsidesi*, *N. parvifrons* and *N. murchisoni*; Fortey and Owens (1978, pp. 267, 276) claimed that Whittingtons's figured material included both *M. selwynii* (at his locs. 1, 2, 5) and *M. murchisoni* (at his locs. 3, 4). According to Wells (1925, p. 494, footnote), Fearnsides collected *N. parvifrons* in the Mawddach Gorge south of Allt Lŵyd and the same form (or its associates?) in the outlier on the summit of Rhobell Fawr (ibid., p. 583).

Farther south, near Hengwrt Uchaf, Salter collected specimens of *M. selwynii*, including the type specimen (Fortey and Owens, 1978, p. 268, pl. 8, fig. 7). The exact locality is uncertain, but the ground west of Hengwrt Uchaf is mapped as volcanic sandstone in the Allt Lŵyd Formation and basic tuff of the Melau Formation.

The age of these 'Henllan Ash' faunas is early in the Arenig epoch, within the *D. extensus* Zone. The faunas from Whittington's localities underlie the *D. hirundo* Zone and, according to Fortey and Owens (1978, p. 236), can be correlated with beds at the top of the Ogof Hên Formation *or* at the base of the Carmarthen Formation in south-west Wales, strata which are also within the *extensus* Zone and well below the *hirundo* Zone. There is no evidence for the presence of the *hirundo* Zone in the present district, and it may be that the Allt Lŵyd Formation extends no higher than the *extensus* Zone. Fearnsides recorded the *hirundo* Zone in his Filltirgerig Beds in the Arenig area, just east of the area studied.

Beds below those mapped as the Benglog Volcanic Formation have yielded sparse graptolite faunas north and north-west of Twr-y-maen: *Amplexograptus confertus*, *Glyptograptus dentatus* and a didymograptid fragment suggest the presence of the Llanvirn Series, *D. bifidus* Zone. In the south-east corner of the district Llanvirn faunas have also been collected below the Benglog Formation. In Nant Helygog, beds below the Brithion Formation yielded *Ogyginus corndensis*. Lewis (1936) recorded '*Trinucleus*' from them, Bates (1965) described *Iocrinus brithdirensis*, and Cox and Lewis (1945, p. 80) mentioned the occurrence of a conulariid and *Didymograptus* of the *bifidus* group. These beds are referred to the *D. bifidus* Zone. In Afon Celynog siltstone and mudstone below the Brithion Formation yielded *A. confertus*, *Didymograptus artus* and *Orthograptus sp.*; Lewis (1936) mentioned also *Cryptograptus tricornis*. These records are evidently of Llanvirn age.

The youngest fossiliferous Ordovician rocks in the present district are apparently dark pyritous mudstone below the Benglog Volcanic Formation in Afon Eiddon, west of Ty-newydd-y-Mynydd. During the present survey these have yielded numerous very poorly preserved diplograptids of doubtful ?post-Arenig age. At this locality, however, Wells (1925, p. 506) obtained *Dicellograptus sextans*, *Dicranograptus rectus*?, *Hallograptus sp.*, and other graptolites no older than the *Nemagraptus gracilis* Zone at the top of the Llandeilo or the base of the Caradoc Series. From Wells' work it is evident that the succession in the Llanvirn–Llandeilo is condensed, at least in the area east of Rhobell Fawr. AWAR

CHAPTER 4

Intrusive rocks

There are numerous intrusions within the Cambrian and Ordovician strata, mostly on the eastern side of the Harlech Dome (Figure 5). Ramsay (1881) divided them into 'feldspathic traps' and 'greenstones'. Andrew (1910), however, divided them into diabase and porphyry. Later, Wells (*in* Matley and Wilson, 1946) recognised the main rock types emplaced within the Cambrian strata as being of basic and intermediate composition, though these divisions did not correspond with those of Ramsay or Andrew. This distinction was not made on Matley and Wilson's (1946) map. They, however, recognised three different forms of intrusion: sills; large, partly transgressive bodies; and dykes. They found only dolerite among the dykes, only intermediate rocks among the large intrusions, but both among the small sills. Emplacement of rocks of both types in all three forms of intrusion preceded the cleavage.

Wells (1925) attributed the intrusions to two magmatic episodes. Developing an idea first proposed by Ramsay (1881), he suggested that many of the intrusive rocks in the Cambrian strata were comagmatic with the Rhobell Volcanic Group and, indeed, he included them in that group. The sills within the Aran Volcanic Group, almost all of which are basic in composition, he believed to be Ordovician in age. Kokelaar (1977, 1979) confirmed the relationship between the Rhobell Volcanic Group and the adjacent intrusions, whereas Allen, Cooper, Fuge and Rea (1976) examined the chemistry of igneous rocks on the eastern side of the Harlech Dome and identified a pre-Arenig episode of sub-alkaline, basic and intermediate intrusions and a chemically distinct group of post-Arenig dolerites. They were also able to produce some evidence of a genetic association between the early magmatism and the porphyry copper mineralisation at Coed-y-Brenin (p. 80). WJR, PMA

GENERAL ACCOUNT

Most of the major intrusions within the Cambrian are laccolithic though some have locally discordant contacts. The majority are microtonalite, but large intrusions of dolerite, quartz-microdiorite and microdiorite occur locally. All these intrusions occur on the eastern side of the Harlech Dome, and are confined to stratigraphic levels within or above the Gamlan Formation, most being high within the Maentwrog or Ffestiniog Flags formations. Besides large-scale doming, the bedding in the strata immediately overlying some intrusions is crumpled and folded. Brecciation is not common. Interfingering of the country rock with intrusive rock around the edges of some bodies, after the fashion of cedar-tree laccoliths, is fairly common. Thermal metamorphic effects, mainly spotting, are confined to aureoles no more than a few metres wide.

Along the western margin of the Rhobell Volcanic Group are three complexes of dolerite and subordinate microdiorite and microtonalite. Ramsay (1881), Wells (1925) and Kokelaar (1977, 1979) regarded these as being co-magmatic with the Rhobell Volcanic Group. In Kokelaar's view these intrusions, which share chemical and mineralogical similarities with the lavas on Rhobell Fawr, mark the site of a major eruptive centre. The intrusions are sheet-like in being concordant or semi-concordant to the steeply dipping to vertical sedimentary country rock, but Kokelaar (1977) regarded them as being dyke-like with respect to the base of the volcanic pile.

Minor concordant or semi-concordant intrusions occur throughout the Cambrian succession. They are mainly dolerite, microdiorite, and quartz-microdiorite with, least commonly, microtonalite and a few bodies of dacite. The dolerite is usually porphyritic but some non-porphyritic ophitic ilmenite-rich dolerite is also present. There appears to be both horizontal and vertical variation in the distribution of these rock-types. Intrusions are most common in the southern and eastern parts of the Harlech Dome at all stratigraphic levels and there is a tendency for dolerite to be dominant in western and north-western parts with an eastward increase in the proportion of intermediate rocks. The vertical distribution of sills appears to be partly influenced by stratigraphy; for example there are few intrusions in the arenaceous Rhinog and Barmouth formations. There is, however, an overriding vertical zonation. The sills within strata below the Rhinog Formation include the majority of the dacites. The greatest concentration of basic and intermediate sills occurs within the Clogau and Maentwrog formations, though they are abundant within the Ffestiniog Flags Formation in areas close to the crop of the Rhobell Volcanic Group. In the Cwmhesgen Formation intrusions are rare except for a suite of very high level dolerite sills immediately beneath the volcanic pile. The intermediate and acid intrusions do not differ petrographically or compositionally from their counterparts among the major laccolithic intrusions and the subvolcanic complexes. The dolerites, however, fall into two distinct petrographic and chemical families (p. 50); one is low in Ti and correlates with the basic rock in the Rhobell Volcanic Group; the other is Ti-rich and is less common, forming the basic intrusions within the Aran Volcanic Group.

Dykes, almost exclusively of dolerite, are most numerous in a swarm trending north-west, but with a subordinate, older NE-trending suite, cutting rocks of the Harlech Grits Group in the central parts of the Harlech Dome. Dykes are less common in rocks above this group, but they intrude all formations including the sills and laccolithic intrusions and the Rhobell Volcanic Group. Within the confines of the district few dykes have been found intruding the Aran Volcanic Group. The dykes are commonly cleaved at the margins. The same two petrographic and chemical types of dolerite are present, though the less common high-titanium variety is confined to NW-trending dykes.

Intrusions within the Aran Volcanic Group are almost entirely sills of dolerite. There is one small body of microtonalite and another, on the margin of the district, of microdiorite.

Over thirty intrusive breccias have been recorded in the district and all lie within a 3 km wide belt which crosses the sheet in a roughly NNE direction from Dolgellau to Moel-y-Slates (Figure 5). With one exception all are situated within the outcrop of the Ffestiniog Flags Formation. Six are pipe-like and several others are of indeterminate form, but the majority are dyke or sill-like and associated with rocks in the Moel y Llan and Nannau complexes (see below). Rice and Sharp (1976) commented on the presence of intrusive breccias in the porphyry copper deposit at Coed-y-Brenin, also within this belt, and there seems to be little doubt that the breccias formed during the late stages of the Rhobell magmatic episode. Autobrecciation, in places with considerable hydrothermal alteration, but not involving transportation of breccia fragments, is also common in the subvolcanic complexes.

All these intrusions have been subjected either to late stage hydrothermal alteration or regional metamorphism or both. The rock names used, however, are those that would apply to the unaltered rock. The majority of the rocks have the mineral composition of gabbro, diorite, quartz diorite and tonalite under the scheme proposed on behalf of the IUGS Subcommission on the Systematics of Igneous Rocks by Streckeisen (1976). Crude modal analyses were done on the acid and intermediate rocks but, because of the likelihood of wrongly attributing the alteration products either to feldspar or ferromagnesian minerals, names for siliceous rocks were chosen mainly on the content of modal quartz rather than on its proportion of the total of light coloured minerals as recommended by Streckeisen (1976). The proportions used are 0 to 5 per cent modal quartz for microdiorite, 5 to 10 per cent for quartz microdiorite and >10 per cent for microtonalite. This classification was tested chemically by Kokelaar (1977) who found that in most rocks, despite albitisation and alteration, the silica content concurred with the petrographic classification adopted.

Following Hatch, Wells and Wells (1956) a median grain size of 1 mm was used to define the boundary between plutonic and hypabyssal and other fine-grained rocks. PMA

PETROGRAPHIC DETAILS

Ilmenite dolerite

This rock and its altered equivalents comprise most of the sills in the Aran Volcanic Group, some minor sills in the Cambrian succession, and some of the dolerite dykes. It is commonly medium-grained, always non-porphyritic (Plate 9.2), and consists of essential plagioclase, augite and ilmenite (~5 per cent) with various alteration products (E 42184, 46070, 47472 are sills, E 42233 from a NW-trending dyke, E 44155 from a dyke in a laccolith). The plagioclase is twinned and normally zoned with labradorite cores, but in many rocks it is albitised and altered to sericite, clinozoisite, chlorite and carbonate. Ophitic augite is commonly partly replaced by colourless or pale green amphibole, and both are replaced at the edges and along cleavages by a fibrous variety. The pyroxene is locally replaced by chlorite. In some rocks small anhedral crystals of deep green or pinkish brown and green, zoned amphibole crystals form a reaction rim to pyroxene crystals. Skeletal ilmenite is the main accessory mineral, in places rimmed by sphene, commonly leucoxenised. Magnetite is a minor mineral. The intergranular mesostases are composed mainly of feldspar, chlorite and quartz with some alteration minerals.

In the coarse-grained varieties from near Brithdir the pegmatitic schlieren consist of feldspar and ilmenite crystals up to 2 cm in length. Augite is rarely ophitic except near the margins. The crystals are tightly packed and feldspar are commonly bent and broken.

In highly altered sills this dolerite (E 43571) consists of totally albitised plagioclase, ilmenite or its pseudomorphs, chlorite and carbonate. Such rocks tend to retain their original textural characteristics, although original augite is difficult to determine.

PND, WJR, PMA

Porphyritic dolerite

Dolerite which varies from sparsely to abundantly porphyritic is the principal basic rock type among the minor sills in the Cambrian, the high-level sills beneath the Rhobell Volcanic Group, dykes in the group and the subvolcanic intrusion complexes, the major basic intrusions and, in highly altered forms, among the dyke swarms. The only fresh forms of the rock are among the high level sills around Rhobell Fawr. Here Kokelaar (1977) recognised the intrusive equivalents of the augite- and plagioclase-phyric basalts that comprise the Rhobell Volcanic Group. These dolerites consist of augite, plagioclase, minor interstitial primary quartz, and secondary minerals that include chlorite, epidote, fibrous amphibole, sericite, calcite, quartz, limonite and pyrite. The augite phenocrysts are subhedral or euhedral, range up to 1 cm in diameter, and commonly are pseudomorphed. They make up no more than 7 per cent of the rock. Plagioclase phenocrysts, however, may be far more abundant. There are numerous limonitic pseudomorphs after magnetite. In rocks with a coarse-grained groundmass augite may be subophitic.

Recognisably similar rocks are present in all other parts of the district, but levels of alteration are always higher. Primary augite is very rarely preserved and it is not always possible to distinguish between altered derivatives of these rocks and amphibole-bearing dolerite.

The Foel Boeth intrusion (see p. 53) is medium-grained, sparsely porphyritic, amygdaloidal and highly altered. Both in phenocrysts and groundmass, plagioclase laths are albitised (E 43133, 41242) and in places replaced by clinozoisite, minor chlorite and calcite. Phenocrysts of ferromagnesian minerals are replaced by chlorite or tremolite, chlorite and epidote. In parts of the intrusion tremolite needles are abundant throughout the rock. Elsewhere replacement by carbonate is intense and tremolite is absent. Amygdales up to 1 cm in diameter contain various combinations of epidote, quartz, chlorite, carbonate, pyrite and sphene. A sample from the chilled margin (E 43135) is fine-grained, light grey, non-porphyritic, and contains feldspar, patchy chlorite, rare sericite and epidote in a mosaic with some quartz. Adjacent medium-grained rocks are altered mainly to carbonate and sericite.

The Moel y Feidiog intrusion (E 41242) (see p. 53) and the dolerite sills to the north of it are petrographically similar to Foel Boeth. In one of the sills on Pen-y-Feidiog [7815 3295] partly chloritised euhedral augite (E 43127) has survived alteration, and relicts of pale green and pinkish brown amphibole crystals also persist. Similar rocks (E 45901, 48509) form sills near Abergwynant quarry [6720 1778] and at Tyddyn Gwladys [7339 2633].

In many of the minor sills and the NW-trending dykes, alteration is sufficiently intense to destroy most primary textural and mineralogical characteristics. Porphyritic texture is in places preserved in the dykes (E 42223), but commonly both sills (E 45892)

and dykes (E 42240) are reduced to granular aggregates of carbonate (mainly calcite with some siderite), chlorite, albite, quartz, magnetite and limonite. Sericitisation is locally intense, whereas both sills and dykes in the central parts of the Harlech Dome are intensely epidotised.

Amphibole-bearing dolerite

Porphyritic dolerite containing pseudomorphs after phenocrysts of both augite and amphibole are said by Kokelaar (1977) to be relatively uncommon among the subvolcanic intrusion complexes and the high level sills around Rhobell Fawr. Euhedral or subhedral phenocrysts of plagioclase, commonly albitised, pseudomorphs after augite (up to 7 mm) and amphibole (up to 2 cm), are set in a fine-grained groundmass mainly of plagioclase laths and limonitic pseudomorphs after magnetite. The rocks show the same type of alteration as the other dolerites. Unaltered primary amphibole is rare. The phenocrysts, which may be embayed, are pseudomorphosed mainly by chlorite. In a sill at Coed Ffridd-arw [7439 1788] there are euhedral phenocrysts (E 46624) of colourless to pale green amphibole, 7 mm in diameter with small patches of optically continuous pinkish brown amphibole. However this mineral apparently replaces other ferromagnesian minerals in other dolerites and its origin is not certain.

Basalt

At Pont y Gain [7510 9273] three white weathered intrusions no more than 15 cm thick pass laterally from dyke to sill and texturally are amygdaloidal basalt (E 41764). Amygdales up to 0.5 mm diameter and crudely oriented pseudomorphs after feldspar phenocrysts up to 2.5 mm are both composed predominantly of calcite and white mica with some albite and quartz. The groundmass comprises altered feldspar microlites in a dusty matrix of carbonate, minor sericite and abundant rod-like opaque (?magnetite) crystals.

Porphyritic microdiorite

This rock occurs predominantly among the minor sills and in the subvolcanic intrusion complexes. It is a component only of the Hafod-y-fedw complex (see p. 56) among the major bodies.

In relatively mildly altered forms it contains euhedral and subhedral, turbid plagioclase phenocrysts up to 4 mm, and rare glomeroporphyritic aggregates and pseudomorphs after amphibole (Plate 10.3), in a fine-grained groundmass with a median grain size about 0.1 mm. The plagioclase phenocrysts are all albitised and either patchily or totally replaced by calcite and sericite. Pseudomorphs after amphibole are usually calcite and fine, granular opaque mineral, or calcite and chlorite. The groundmass is composed mainly of feldspar with rare quartz. Textural variations range from granular plagioclase showing anhedral intergrowths (E 45905) to strongly flow-oriented plagioclase laths (E 48526). In some very thin sills trachytic texture was observed (E 41755). Chlorite is present in a few rocks. Pyrite commonly concentrates at the sill margins (E 41767) but may be present locally as 1-cm cubes. Accessory minerals include apatite, zircon, sphene and ilmenite. Varieties of this rock from within the Dolwen Formation are distinctive in the abundance of chlorite in the groundmass and the presence of epidote in pseudomorphs after both plagioclase and amphibole (E 44165, E 54035).

The majority of these rocks are highly altered. The phenocrysts of feldspar are replaced by sericite and carbonate, and the amphibole phenocrysts by carbonate and chlorite with or without limonitised opaque minerals. Groundmass characteristics are largely obscured by pervasive carbonate (calcite and some siderite) replacement (E 44755), or less commonly sericite or chlorite.

Quartz-microdiorite

This rock forms minor sills and is present in the subvolcanic intrusion complexes. It is a component only of the Afon Wen, Hafod-y-fedw and Dol-fawr major intrusion complexes.

Among the minor intrusions the quartz-microdiorite is always porphyritic and highly altered, showing the same main characteristics as the porphyritic microdiorite. Quartz is present as sporadic resorbed, usually recrystallised, phenocrysts (E 46058) or as an interstitial mineral in the groundmass (E 45904). Some rocks display a distinctive, platy recrystallisation texture in the groundmass (E 41980). In general the groundmass is finer grained than in the microdiorite. In a few samples (E 41978) the phenocrysts are so crowded together that the rock appears to be non-porphyritic in hand specimen.

Alteration is less severe in the large intrusions. Primary plagioclase showing normal and oscillatory zoning is preserved with andesine cores in some rocks. The rock tends to be coarser-grained than in the thin sills (E 48494–5, E 48499–501). There is a generation of large phenocrysts of plagioclase and pseudomorphs after amphibole, and a second generation of plagioclase phenocrysts with a median size of about 1 mm which in some rocks are closely packed together and intergrown. The matrix is composed of small feldspar crystals, quartz, chlorite and epidote. PMA, WJR, BPK

Porphyritic microtonalite

This and the sparsely porphyritic variety are dominant among the major laccolithic intrusions in the Cambrian. Microtonalite is the least common rock type among the minor sills and subvolcanic complexes.

The rock varies from grey through greenish grey to pinkish brown. It contains phenocrysts of plagioclase, amphibole, rare biotite and quartz in a matrix of variable grain size mainly composed of feldspar, quartz and chlorite. Quartz rarely exceeds about 17 per cent modal content.

The plagioclase phenocrysts, up to 1 cm long, are euhedral or subhedral, and commonly altered to albite and chlorite, calcite, sericite and quartz. In some rocks there is total replacement by sericite and calcite. Most crystals show multiple twinning. Normal and oscillatory zoning are common (Plate 10.5), and andesine cores (E 47480) have been identified in unaltered rocks. In altered rocks these structures are relict. Some of the phenocrysts are intergrown with the groundmass at the rim. In most rocks the amphibole phenocrysts, up to 1.5 cm long, are represented by euhedral pseudomorphs of varying proportions of chlorite, epidote and calcite, with tremolite, sericite, quartz and sphene as less common alteration minerals. Many pseudomorphs possess a rim of a fine-grained granular mineral of the epidote family. The chlorite shows a wide range of colours and anomalous birefringence. Where primary hornblende (Plate 10.6) has been observed (E 43139) it is greenish brown, commonly with a pale brown core. Quartz phenocrysts (E 41770) are never common and in some intrusions have not been recorded. They are subhedral to euhedral, in places totally recrystallised, and in size are smaller than the other phenocrysts.

Typically, the groundmass (E 44761, E 48307) is an equigranular mosaic of feldspar, quartz, chlorite and epidote with a 0.02 to 0.06 mm grain size range. Less commonly (E 41974, E 44762, E 48308) it is composed of closely packed laths or tabular crystals of plagioclase usually in the order of 0.05 to 0.1 mm, but in rare cases reaching nearly 1 mm. The plagioclase is usually albitised and clouded by clinozoisite and sericite. Only rarely does the plagioclase of the matrix show a preferred orientation. Quartz, chlorite, rare epidote and muscovite are interstitial to the feldspar in rocks with coarser-grained matrix, and in some instances there is a generation of smaller albite crystals between the large ones. K-feldspar has not been recorded though myrmekite has been seen in one rock.

Plate 10
Photomicrographs of intrusive rocks of the Harlech Dome.

1 Intrusive breccia from a thin sill near Llyn Cynwch composed of quartz wacke (light coloured) and muddy siltstone fragments. (E 47481, plane polarised light, ×25).
2 Albitised dolerite from Foel Boeth intrusion with albite laths (ab), tremolite (tr), chlorite (ch) and epidote (ep). (E 43133, crossed polarisers, ×25).
3 Porphyritic microdiorite with plagioclase phenocrysts and chlorite-calcite pseudomorph after amphibole (outlined). (E 48480, crossed polarisers, ×25).
4 Microtonalite from Moel y Slates intrusion showing sericitised plagioclase, interstitial quartz and minor chlorite (dark patches). (E 44153, crossed polarisers, ×30).
5 Porphyritic microtonalite from Afon Wen complex showing oscillatory zoning in andesine. (E 47480, crossed polarisers, ×25).
6 Porphyritic microtonalite from Blaen Lliw intrusion showing partly altered hornblende and totally sericitised plagioclase phenocrysts. (E 48297, crossed polarisers, ×25).

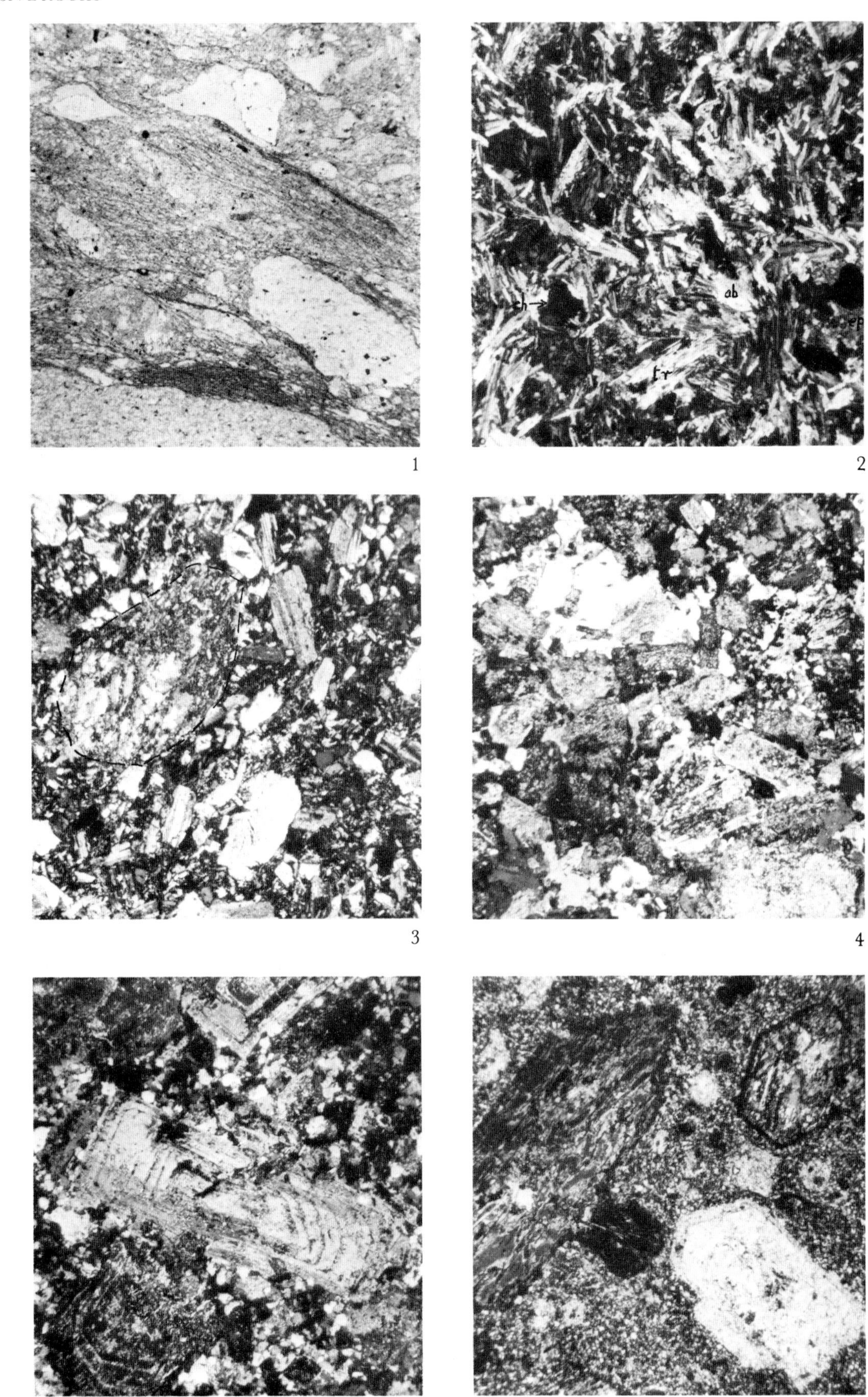

Rarely, the chlorite in the matrix appears to have replaced biotite. In some rocks sericite and calcite are present as alteration products throughout the matrix.

Among the accessory minerals leucoxenised opaques and pyrite are commonest, with chalcopyrite, zircon, apatite and sphene.

Microtonalite

This rock is usually grey or greenish grey with rare phenocrysts (E 48505, 48506, 44793) of plagioclase, usually less than 5 mm long, and slightly smaller amphibole pseudomorphs showing similar form and alteration to those in the porphyritic microtonalite. Euhedral quartz phenocrysts are very rare (E 46236).

The main body of the rock is composed of randomly orientated tabular plagioclase crystals (Plate 10.4), generally in the order of 0.5 to 0.75 mm long but locally with a median size of around 1 mm. Quartz, chlorite and rare epidote are interstitial, and in places myrmekite is present. The plagioclase is commonly altered to albite with chlorite or clinozoisite, calcite and sericite; and in some rocks entirely to sericite. Rarely (E 44153) the plagioclase is only partially altered and retains its primary normal zoning from calcic andesine or labradorite to albite, the latter clear and intergrown with the matrix. Chlorite is abundant in some rocks and possibly replaces both amphibole and biotite. Among the accessory minerals are pyrite, leucoxenised ilmenite, sphene, apatite and allanite.

The Coed Ty-cerrig intrusion [723 248] is texturally distinctive. The rock (E 49445) contains a few plagioclase phenocrysts 0.4 to 1 mm long, but mostly comprises large plates of quartz within which are embedded randomly oriented, altered plagioclase laths about 0.1 mm long. Slender plates of muscovite, minor calcite, stilpnomelane and limonitic pseudomorphs are present with a little chlorite possibly after biotite. PMA, WJR

Dacite

At Pont-y-Gain [7509 3269] and on Moel Oernant [7477 3406] there are thin sills of sparsely porphyritic, amygdaloidal dacite (E 41754, 41765). They are both highly altered to carbonate, minor chlorite and limonite, and are composed of crudely orientated feldspar and interstitial or irregular shaped subophitic quartz. PMA

A distinctive group of quartz-rich acid minor intrusives occurs in the central part of the dome within the Dolwen and Llanbedr formations. These rocks, which weather to a light-coloured crust, are apparently the equivalents of Ramsay's (1881) 'felspathic traps' and of Gilbey's (1969) 'dacites'. They consist of fine- to medium-grained, equigranular, highly leucocratic aggregates of quartz, albitised and epidotised plagioclase and possibly orthoclase. WJR

CHEMISTRY

Representatives of all the groups distinguished above on the basis of different petrography and/or form of intrusion have been analysed. The available data include previously presented analyses of major dolerite and microtonalite intrusions (Allen and others, 1976), intrusion complexes associated with the Rhobell Fawr Volcanic centre (Kokelaar, 1977), and dolerites within the Aran Volcanic Group around the Harlech Dome (Allen and others, 1976) including those of the SW Aran Mountains (Dunkley, 1978). New analyses of the minor concordant intrusions and dykes intruded into the Cambrian, and of rocks of the Afon Wen intrusion complex associated with the Coed y Brenin copper deposit are included. In the new analyses, SiO_2, Al_2O_3, TiO_2, total Fe, P_2O_5, Mo, F and Cl were determined by automated photometric methods; MgO, CaO, MnO, Zn and Cu by atomic absorption spectrophotometry; K_2O and Na_2O by flame emission spectrometry; Ba, Rb, Sr, Y and Zr by XRF analysis. H_2O, CO_2 and S were not determined and this is believed to account for the low totals in some of the new analyses. Average and representative new analyses are presented in Table 8, together with comparative data from the previously presented work. Figures 12 and 13 are Ti-Zr and FMA variation diagrams based on all the available data.

The chemical data presented here conform to the groupings defined previously on field and petrographic criteria. Allen and others (1976) showed that Si, Ti, total iron, Mn, Ca, Mg and Zr were good chemical discriminants for groups of intrusive rocks from the Harlech Dome. Of these the most useful in respect of the new data are Ti, Zr, Si and total iron. The first two elements, being relatively immobile under conditions of low grade metamorphism and hydrothermal alteration (Pearce and Cann, 1973), were expected to be useful but, in view of the close Ti–Si and Zr–Si correlation, Si is also a useful discriminant and appears to have been relatively unaffected by post-emplacement changes. Using Ti and Si, all the intrusive rocks of the Harlech Dome can be assigned either to a Ti-rich group of predominantly basic

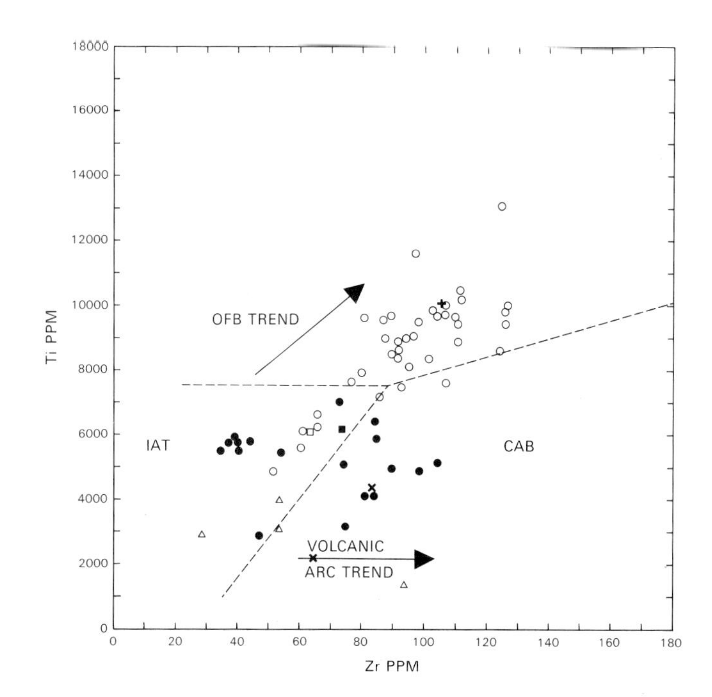

△ Coed y Brenin basic intrusions.
○ Ophitic dolerite intrusions of the SW Arans (Dunkley, 1978).
● Dolerites from the Moel y Llan and Cerniau intrusive complexes, Rhobell Fawr (Kokelaar, 1977).
+ Average dolerite intrusion into Ordovician rocks around the Harlech Dome (Allen et al, 1976).
□ Average albitised dolerite intrusion into Cambrian (Allen et al, 1976).
Field boundaries and trends for ocean floor basalts (OFB) and volcanic arc basalts - island arc tholeiites (IAT) and calc-alkaline basalts (CAB) from Garcia (1978) after Pearce and Cann (1973).
■ Dykes: Low-Ti dolerite.
✕ Minor concordant intrusions: Low-Ti dolerite.

Figure 12 Ti versus Zr for basic intrusive rocks in Harlech Dome and surrounding areas

composition and probably Ordovician in age, or to an older Ti-poor group which includes basic, intermediate and probably also acid compositions.

High-titanium dolerite intrusions

The chemical characteristics of this group include high TiO_2 and total iron and relatively low Al_2O_3. The group includes a few of the minor concordant intrusions from various parts of the dome (Table 8, no. 1) and a number of the NW-trending dykes (Table 8, no. 8). The 'Afon Wen diorites' of Rice and Sharp (1976; Table 8, no. 5) may also belong to this group. More certain, because of very close chemical similarities, is the inclusion in this group of the dolerite intrusions of the SW Arans (Table 8, no. 11) and the dolerites intruded into the Aran Volcanic Group around the S, E and N flanks of the Harlech Dome (Table 8, no. 10). Some of the latter have higher Na_2O contents than the high-titanium dolerites within the Cambrian and may have undergone selective soda metasomatism (Allen and others, 1976).

On the FMA diagram (Figure 13) rocks of this group tend to plot within the tholeiitic field of Irvine and Baragar (1971), and more differentiated varieties such as Dunkley's (1978) pegmatitic dolerites and diorites show moderate iron enrichment. On the Ti–Zr plot (Figure 12) the ophitic dolerites of the SW Arans follow the ocean floor basalt trend, and the TiO_2 content of the high-titanium dolerites from the Harlech Dome is such that these must also plot well within the ocean floor basalt field. Dunkley (1978) considered that the chemistry of the igneous rocks of the SW Arans is compatible with an origin within a marginal basin environment.

Low-titanium intrusions

Most of the minor dolerite intrusions of the Harlech Dome differ markedly in chemistry from those described above.

Table 8 Average representative analyses of intrusive rocks from the Harlech Dome

	CONCORDANT INTRUSIONS				COED Y BRENIN			DYKES		COMPARATIVE DATA					
%	1	2	3	4	5	6	7	8	9	10	11	12	13	14	15
SiO_2	46.3	48.3	59.5	69.5	42.9	56.3	58.5	44.8	46.1	48.0	49.54	46.8	46.9	57.5	56.4
TiO_2	1.86	0.60	0.40	0.15	1.31	0.47	0.40	1.96	0.92	1.68	1.51	1.03	0.90	0.51	0.53
Al_2O_3	13.8	15.5	17.6	15.9	14.7	16.6	15.7	15.8	15.8	14.2	15.14	14.9	16.6	15.7	16.4
Fe_2O_3	3.39	1.46	2.48	1.17	1.10	1.96	1.12	2.83	2.57	11.57	1.73	9.43	2.65	6.24	2.04
FeO	8.98	6.84	2.87	1.10	7.94	3.49	3.49	9.43	7.02		8.26		6.44		4.21
MnO	0.21	0.28	0.18	0.09	0.22	0.13	0.13	0.18	0.20	0.23	0.20	0.17	0.20	0.18	0.20
MgO	6.30	7.37	1.80	0.84	8.83	3.95	4.14	6.94	7.46	7.60	7.19	8.05	5.95	3.47	2.96
CaO	8.20	7.01	6.83	2.09	8.92	3.83	3.35	8.06	8.40	8.93	8.88	8.01	7.67	4.85	5.59
Na_2O	1.92	3.55	4.81	4.94	1.79	1.39	2.92	2.76	2.95	3.50	3.47	3.81	2.70	3.67	3.25
K_2O	0.46	0.40	0.65	1.53	1.04	2.67	2.38	0.46	0.55	0.42	0.68	0.86	0.90	1.47	1.54
P_2O_5	0.25	0.19	0.19	0.13	0.16	0.14	0.12	0.30	0.13	0.19	0.15	0.26	0.15	0.16	0.17
ppm															
F	696	541	359	242	263	455	359	374	399				284		252
Cl	74	103	64	50	30	122	81	87	77				83		50
Cu	23	57	8	4	39	2041	531	33	71	64	61	60	72	57	24
Zn	185	639	72	31	67	40	36	99	861	72	118	60	145	92	171
Rb		11		46		94	58		18		16		36		58
Ba		214		632		243	304		161	166	305	151	319	231	394
Sr		333		327		59	198		325		305		351		276
Y		14		15		15	8		24		30		16		16
Zr		75		184		85	87		74	106	128	64	68	84	97
Mo						17.6	3.3			0.5		0.5			1.4
As						49	5			7		16			9

Key to analyses

1 Minor concordant intrusions: Average high-Ti dolerite.
2 Minor concordant intrusions: Average low-Ti dolerite.
3 Minor concordant intrusions: Average microdiorite to microtonalite suite.
4 Minor concordant intrusions: Average dacite from Dolwen and Llanbedr formations.
5 Coed y Brenin intrusions: Example of high-Ti dolerite ('Afon Wen Diorite' of Rice and Sharp, 1976).
6 Coed y Brenin intrusions: Average 'Older Diorite' (mainly microdiorite and quartz-microdiorite).
7 Coed y Brenin intrusions: Average microtonalite ('Porphyritic Diorite' of Rice and Sharp, 1976).
8 Dykes: Average high-Ti dolerite.
9 Dykes: Average low-Ti dolerite.
10 Average dolerite intrusion into Ordovician rocks around the Harlech Dome (Allen and others, 1976, table 2).
11 Average ophitic dolerite, SW Arans (Dunkley, 1978).
12 Average albitised dolerite intrusion into Cambrian rocks (Allen and others, 1976, table 1).
13 Average dolerite from Moel y Llan and Cerniau intrusion complexes, Rhobell Fawr (Kokelaar, 1977).
14 Average microdiorite, quartz-microdiorite and microtonalite intrusion into Cambrian rocks (Allen and others, 1976, table 2).
15 Average microdiorite, quartz-microdiorite, Rhobell Fawr (Kokelaar, 1977).

NOTE: Analysts for 1–9, R. Fuge and H. Edwards

They have much lower TiO_2 and iron, higher Al_2O_3, and low contents of K_2O and trace elements (Table 8, no. 2). Chemically, they form a relatively homogeneous group although some of the sills and dykes in the south-west (e.g. E 40819) and north-west (E 42305) contain very high Zn (0.18 to 0.34 per cent) and one sill (E 40823) contains more than 1.0 per cent MnO. Low-titanium dolerite is the predominant basic composition of the intrusions of the Harlech Dome and is virtually the only composition found in intrusions on the west side of the dome. The north-easterly and some of the north-westerly trending dykes have a very similar composition (Table 8, no. 9), and the dolerites from the Moel y Llan and Cerniau intrusion complexes (Table 8, no. 13) and the Rhobell Volcanic Group basalts (Kokelaar, 1977) are also similar. The albitised dolerites (Table 8, No. 12) of Allen and others (1976) are also thought to belong to this group although they contain slightly higher TiO_2 and higher K_2O. The latter feature may be explained by the fact that some of the rocks analysed by Allen and others (1976) have undergone a potash metasomatism associated with porphyry copper mineralisation in the south-east of the area.

Analyses of the minor concordant microdiorite, quartz-microdiorite and microtonalite intrusions (Table 8, no. 3) show close similarities with intermediate intrusive rocks from Rhobell Fawr (Table 8, no. 15) and large intrusions into the Cambrian (Table 8, no. 14). All of these rocks lie on a continuation of the chemical trends shown by the low-titanium dolerites and share with them chemical properties such as high contents of Al_2O_3 and low TiO_2, K_2O and immobile trace elements. It is considered, therefore, that the low-titanium dolerites and these intermediate intrusive rocks form part of the same suite. It is also possible that the minor intrusions of dacite intruded into the early Cambrian near the core of the dome (Table 8, no. 4) represent a further continuation of this compositional trend. Although there is a gap between the SiO_2 content of these intrusions (SiO_2 = 69 per cent) and those of the microtonalites (maximum SiO_2, 64 per cent), characteristics such as low Ti and high Al_2O_3 suggest that the 'dacites' may be related to the intermediate rocks.

Chemically the 'Porphyritic Diorites' of Rice and Sharp (1976) in the Coed y Brennin mineral deposit differ from microtonalites elsewhere on the dome (compare Table 8, nos. 7 and 3) in having lower CaO, Na_2O and higher K_2O and Cu. However, similarities in petrography and contents of immobile elements such as Ti and Zr, and FMA trends (Figure 13) suggest the 'Porphyritic Diorites' are the equivalents of the Harlech Dome microtonalites but have undergone copper mineralisation and alteration, including K_2O metasomatism subsequent to their emplacement. Chemically, their 'Older Diorites' (Table 8, no. 6) form a rather heterogeneous group including some basic rocks which may be transitional from microdiorites to low-titanium dolerites. Generally speaking the 'Older Diorites' contain less SiO_2 than the 'Porphyritic Diorites', but the close resemblance of Ti and Zr suggests a genetic relationship between the two groups. The differences that exist, e.g. higher values of K, Mo, Cu, As in the 'Older Diorites', can be explained by the effects of a more intense and/or prolonged hydrothermal mineralisation than that which appears to have affected the younger 'Porphyritic Diorites'. Compared with the intermediate intrusions of the Harlech Dome, the 'Older Diorites' show similar immobile element contents but they contain much greater amounts of K_2O and Cu, and less CaO, Na_2O, MnO and Zn. Chlorine is also enriched, particularly in rocks showing phyllic alteration. Again, these differences can be attributed to the effects of mineralisation and hydrothermal alteration, especially of phyllic type at Coed y Brenin. Unlike most intrusive rocks from elsewhere in the district, the Coed y Brenin rocks plot on the potash-rich side of the normal igneous alkali spectrum (Hughes, 1972) and K/Ba ratios (91 in average 'Older Diorite') are much higher than in average Harlech Dome intrusive rocks, suggesting introduction of potash (Allen and others, 1976, p. 106).

The low-titanium intrusion group is, therefore, considered to embrace not just low-titanium dolerites, but also microdiorites and microtonalites, including those of Coed y Brenin. On an FMA diagram (Figure 13) these rocks plot almost exclusively in the calc-alkaline field of Irvine and Baragar (1971) and show little if any iron enrichment. All

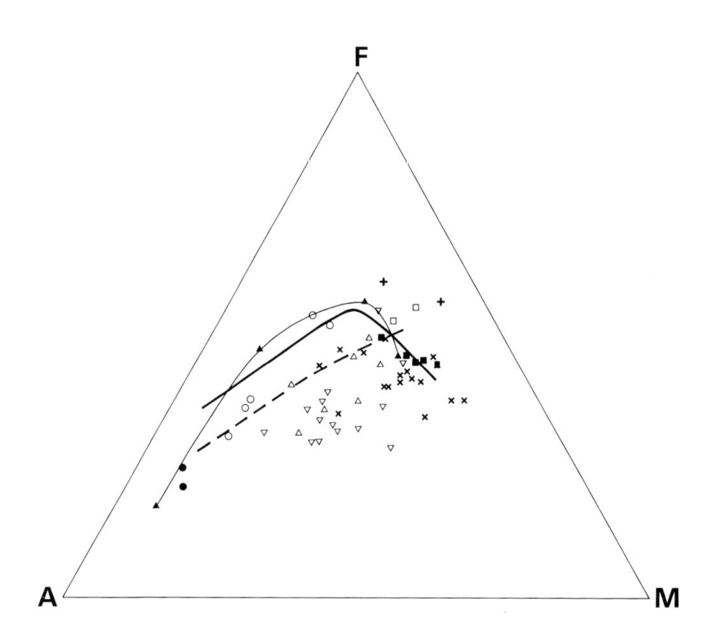

+ Minor concordant intrusions: High-Ti dolerite.
× Minor concordant intrusions: Low-Ti dolerite.
○ Minor concordant intrusions: Microdiorite and microtonalite.
● Minor concordant intrusions: Dacite.
△ Coed y Brenin intrusions: 'Older Diorite' (mainly microdiorite and quartz microdiorite).
▽ Coed y Brenin intrusions: Microtonalite ('Porphyritic Diorite').
□ Dykes: High-Ti dolerite.
■ Dykes: Low-Ti dolerite.
—— Boundary between tholeiitic (above) and calc-alkaline (below) fields (Irvine and Baragar, 1971).
—▲—▲— Trend for intrusive igneous rocks of the SW Arans. Symbols show average ophitic dolerite, pegmatitic dolerite, diorite and intrusive rhyolite (Dunkley, 1978).
– – – Trend for Moel y Llan and Cerniau intrusive complexes, Rhobell Fawr (Kokelaar, 1977).

Figure 13 FMA diagram for intrusive rocks in the Harlech Dome and surrounding areas

the rocks plot in the fields of island-arc tholeiite or calc-alkaline basalt on the Ti–Zr diagram (Figure 12), following a typical volcanic arc trend. All these chemical properties define the group as calc-alkaline, and various chemical characteristics such as relatively low K_2O and residual trace element contents favour an island-arc rather than a continental margin setting (Allen and others, 1976). WJR

AGE RELATIONS, CORRELATION AND ASSOCIATIONS

The petrographic, chemical and field data available suggest that there are two major magmatic episodes represented by these rocks, both related to volcanism and predating the regional cleavage. All the major intrusion complexes, the three subvolcanic complexes, most of the minor concordant intrusions in the Cambrian and most of the dyke rocks belong to the late-Tremadoc Rhobell magmatism, and were emplaced after folding that was gentle in a regional sense but more intense around Rhobell Fawr. The derivatives of this magmatism include the low-titanium porphyritic dolerites, amphibole-bearing dolerites, microdiorite to microtonalite suite, and possibly the dacitic intrusions low in the Harlech Grits Group. Kokelaar (1977) showed that the microdiorite to microtonalite series of intrusions post-date the dolerites in the Moel y Llan and Cerniau complexes. Low-titanium altered dolerite dykes, however, intersect minor sills of all compositions and microtonalite laccoliths (e.g. Dôl-haidd) and microdiorite sills in the Nannau complex; it is suggested that the major phase of regional dyke emplacement took place in the rising Harlech Dome towards the end of the Rhobell magmatic episode. Though the dykes trend mainly north-west, intrusion took place also along a complementary set of (?conjugate) north-east fractures.

The second magmatic episode gave rise to the high-titanium ophitic augite dolerite intrusions which comprise some of the high-level sills in the Cambrian and the microdiorite intrusion in the Aran Volcanic Group in the south-east corner of the district. Davies (1959) observed the Cader Idris granophyre to intrude a dolerite sill of this family on Cader Idris, and Dunkley (1978) recorded no sills higher than low in the Aran Fawddwy Formation. The evidence of wet sediment deformation associated with the intrusion of these sills (Dunkley, 1978) and the chemical similarity observed by Allen and others (1976) with the basalts in the Aran Volcanic Group suggest that this phase of intrusion was more or less contemporaneous with eruption of the latter. Dunkley (1978) considered that the suite of intrusions in the Aran Volcanic Group is tholeiitic in character, and observed in the Brithdir intrusions FeO (total) and TiO_2 enrichment during early and middle stages of differentiation, the peak of iron enrichment being represented in the late-stage pegmatitic schlieren and veins. Analyses presented here confirm the tholeiitic character of the high-titanium dolerite suite which shows many chemical similarities to ocean floor basalts. Although there are many uncertainties, a transition from an island arc setting in the late-Tremadoc Rhobell igneous episode to a marginal basin setting in the Ordovician is consistent with the chemistry. WRJ, PMA

DETAILS OF INTRUSIONS

The forms of intrusion, and the composition, petrography and magmatic affinity of the intrusive rocks display considerable diversity. No single, simple classification is adequate. In this section the intrusions are classified largely on their form and occurrence under the following main headings: major intrusions in the Cambrian; subvolcanic intrusion complexes; minor concordant intrusions within the Cambrian strata; dykes; intrusions within the Aran Volcanic Group; intrusive breccias.

Major intrusions in the Cambrian

The main concentration of microtonalite intrusions is in the north-east of the district around and to the east of Nant Braich-y-ceunant. The laccoliths, all within the Maentwrog and Ffestiniog Flags formations, probably form an interconnected series which extends northwards beyond the district. They are, however, traversed by a number of strong faults; interrelationships are not always clear. The **Craiglaseithin** and the smaller **Nant Hîr** intrusions to the south-west of Nant Braich-y-ceunant are the only other major microtonalite intrusions in the north-east of the district.

To the west of Rhobell Fawr, major concordant intrusions situated stratigraphically below the high level subvolcanic dyke complexes are more complex than in the north-east and include quartz-microdiorite and microdiorite components in addition to varieties of microtonalite. Among these intrusions are those at **Hafod Fraith, Coed Ty-Cerrig, Afon Wen, Hafod-y-Fedw** and **Dôl Fawr.**

Major basic intrusions include only the **Foel Boeth** laccolith and **Moel-y-Feidiog sill**, both emplaced within the Ffestiniog Flags Formation, and the higher-level **Bryn Brâs** laccolith.

Foel Boeth

The intrusion of dolerite on Foel Boeth, its probable continuation on Pen-y-Feidiog, the suite of sills to the south-west of Siglen-las [781 329], all situated within the Ffestiniog Flags Formation, and a dyke west of Moel-y-Feidiog [7782 3242], which also intrudes the Maentwrog Formation, are compositionally similar and show the same style of alteration (p. 53). Contacts exposed on all sides of the Foel Boeth intrusion show that it lies within the core of an anticline which may predate the intrusion. Except for a small upward protrusion from the main body on the western side, only the roof is exposed. The contact is visible at the northern end [783 348] where the uppermost 20 cm are chilled and flow banded. Elsewhere [7838 3383] the banding is folded. Sedimentary rocks on top of the intrusion are hornfelsed, tightly folded, locally brecciated and veined by dolerite and quartz. Some xenoliths of local roof-rock occur within the dolerite.

The main rock type (Plate 10.2) is greenish grey, sparsely porphyritic, medium-grained, amygdaloidal albitised dolerite (E 43133) with abundant secondary fibrous amphibole. In places the rock is pyritic and mildly altered to carbonate (E 43126) and contains no amphibole, but all along the western margin there is a zone of intensively altered rock almost completely converted to carbonate, sericite and quartz (E 41983).

Moel-y-Feidiog

The intrusion is folded and intersected by a N–S fault. On the east of it the exposed contacts [eg. 7852 3142] suggest that only the upper surface of the intrusion in the core of an anticline is exposed. On the west of the fault, however, there is a basal contact [7795 3209] and the intrusion has the form of a southward dipping sheet containing large enclaves of Ffestiniog Flags Formation. On both sides of the fault the contact is locally brecciated, and in places [7851 3145] there is a mixing of dolerite and country rock fragments. The main rock type (E 41242) on both sides of the fault is sparsely porphyritic, amygdaloidal, light grey albitised dolerite. The rock is patchily altered to calcite and, as at Foel Boeth, there is a marginal phase highly altered to carbonate. In places, however, the rock contains much epidote and chlorite (E 42420). A local variety is distinctive in containing plentiful feldspar phenocrysts (E 42491). PMA

Bryn Brâs laccolith

This laccolith, of leucocratic quartz-dolerite [7805 2228], intrudes the Rhobell Volcanic Group and lies close to the westward dipping basal contact of the group. In the north-east and south-west the contact is steep where it transgresses the subvolcanic unconformity. In the west and south-east the intrusion was concordantly emplaced. Though Wells (1925, p. 473) described an intrusive contact in the north-east he did not recognise the discrete form of this intrusion.

Remnants of the domed upper surface of the intrusion have been uncovered around much of the outcrop [7754 2228]. Xenoliths derived from the volcanic host are common close to the contact and others, including coarse grit, occur sparsely. Breccias near the northern margin of the intrusion may represent a roof pendant [7775 2308]. Chilled margins approximately 1 cm wide have been recorded.

The zone of contact metamorphism in the sedimentary rocks varies from 10 m in the north-east [7818 2290] to a few centimetres in the south. The relatively extreme thermal effect in the north-east is attributed to prolonged flow of magma close to the contact. If the intrusion is reorientated so that the base of the volcanic group is close to horizontal, the contact at which hornfelsing is pronounced becomes overturned, while elsewhere the contacts dip away from the middle of the intrusion. It is possible, therefore, that the north-east contact is that of a feeder to the intrusion.

Well developed cooling-contraction joints occur throughout the outcrop and result in striking terrace-like features on the steeper slopes.

Unlike most other intrusions around Rhobell Fawr the laccolith underwent differentiation *in situ*. Mineralogical and textural layering have not been recognised but there is a chemical variation with level in the exposed portion. The deepest exposures, in the south-west, are more basic than others, and it is thought that the more ferromagnesian-rich part of the intrusion is unexposed. The marginal facies is darker coloured and is probably undifferentiated dolerite. BPK

Braich-y-ceunant

The interlinked complex of intrusions around Nant Braich-y-ceunant (Figure 14) is distributed through a stratigraphic thickness of over 1 km. Stratigraphically, the Pen-y-rhôs, Dôl-haidd, Glasgoed and Range intrusions form the lowest part of the complex (Figure 15) with those of Moel-y-Slates, Bwlch-y-Bi and Waen Blaen-lliw above them.

The somewhat asymmetric **Range** intrusion is laccolithic in form. The top can be traced north-eastwards along Afon Gain [from 7627 3460] and is exposed in the stream [7636 3477]. The contact is sharp and steeply dipping. A small body of intrusive breccia of indeterminate shape lies adjacent to the contact [7659 3488]. Near Buarth-brwynog [7587 3477] the dip of 78° SSW in sedimentary rocks at the contact converts to an approximate NNE dip when a regional eastward dip of anything greater than 30° is returned to the horizontal, suggesting that this may be the base of the intrusion. The intrusion consists almost entirely of porphyritic microtonalite, but near Buarth-brwynog a small body of porphyritic microdiorite (E 41976) intrudes the microtonalite and contains xenoliths of it.

The **Dôl-haidd** intrusion of porphyritic microtonalite is separated from the Range intrusion at its attenuated northern end by a fault. Its base appears progressively to cut across older strata northwards. Two thin sills of similar rock types underlying the intrusion are exposed in Afon Prysor [c.7602 3692]. At Castell Prysor [7578 3687] an oval area of porphyritic microtonalite, xenolithic at the margin [7575 3672], within tightly-folded country rock, though not itself deformed, is interpreted as a possible feeder to the Dôl-haidd intrusion. The top of this intrusion is not clearly defined, but in the bed of Nant Braich-y-ceunant [around 7672 3613] there are several bodies of porphyritic microtonalite, some discordant, within hornfelsed country rock immediately above the main body of the intrusion. Contacts are usually brecciated, and it is not certain whether this is a breccia zone on the roof of the main intrusion or a zone of minor intrusions. The eastern termination of the upper part of this intrusion is thought to lie east of a major fault [772 362] where a thin sill protrudes from the apparently discordant upper surface. Dykes of highly altered dolerite trending roughly north-west and fresh augite dolerite trending west-north-west cut this intrusion.

The **Glasgoed** intrusion of porphyritic microtonalite, which lies in heavily drift-covered ground west of the Bryn-celynog fault, is considered to be a westward projection from the Dôl-haidd intrusion.

On the eastern side of Moel Uchaf Dôl-haidd a group of NW-trending faults separates the Dôl-haidd and **Pen-y-rhôs** intrusions. Within the fault zone there are several sheet-like intrusions, some xenolithic, in a sequence of hornfelsed interbedded mudstone and coarse quartzose siltstone low in the Maentwrog Formation. Those in the northern part of the zone are presumed to be protrusions from the lowest part of the Pen-y-rhôs body, which itself extends north of the sheet splitting into a number of sheet-like, locally anastomosing components.

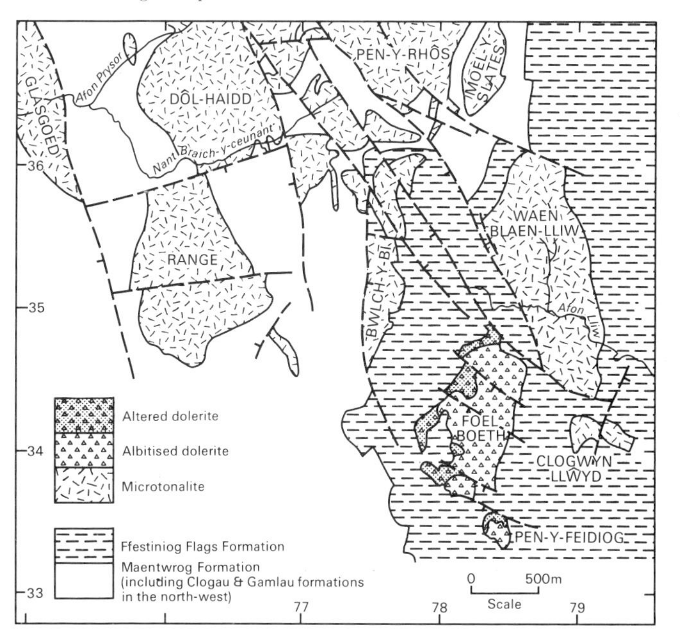

Figure 14 The intrusive complex of Braich-y-ceunant and nearby dolerite intrusions

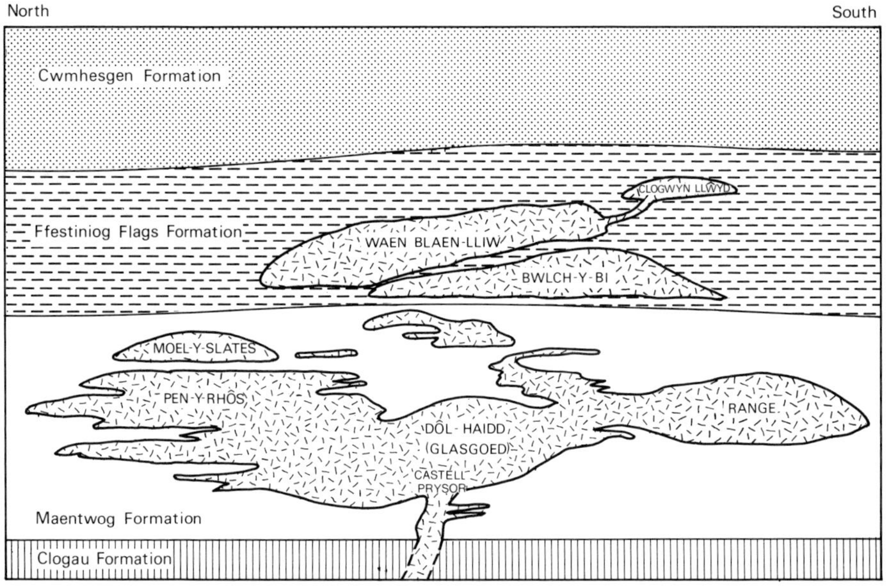

Figure 15 Diagrammatic reconstruction of a N–S section through the Braich-y-ceunant complex before the main folding and faulting

The **Moel-y-Slates** laccolith, exposed on the westward facing scree-covered scarp of Moel-y-Slates, is 800 m long and no more than 200 m thick in the middle parts. It was intruded into the upper part of the Maentwrog Formation, above the Pen-y-rhôs intrusion. Sedimentary rocks above the upper contact are baked and tightly folded. The intrusion, unlike those below it, consists of uniform light grey, even grained, sparsely porphyritic microtonalite (E 44152–3).

The sparsely porphyritic microtonalitic **Bwlch-y-Bi** intrusion (E 46348, 45236), similar to Moel-y-Slates, is about 200 m thick and was intruded just above the base of the Ffestiniog Flags Formation. The main exposure, intersected by faults, is in a westward-facing escarpment largely covered in scree and blocky head. In the presumed extension of this intrusion across faults [7766 3601] the base is perfectly conformable. A sill of intrusive breccia no more than 60 cm thick (p. 59) is present along the top of the intrusion [7774 3510]. Three others are present hereabouts, one following the base of a 9-m sill of porphyritic dacite about 6 m above the Bwlch-y-Bi intrusion.

The eastward-dipping intrusion of porphyritic microtonalite at **Waen Blaen-lliw** measures 350 m across its thickest part and intrudes the Ffestiniog Flags Formation near its base; it is at a similar level to that at Bwlch-y-Bi, but is separated from it by a NW fault zone. The lithologies in the two intrusions, however, are quite different. The base of the Waen Blaen-lliw intrusion is in the northern part of the exposure; the top appears to rise stratigraphically southwards. It is likely that this intrusion is linked to the small Clogwyn Llŵyd body by a discordant feeder exposed between the two [7903 3435]. The intrusion displays excellent columnar jointing. This is the only intrusion in this area known to contain fresh hornblende (E 43583).

Craiglaseithin

This intrusion of porphyritic microtonalite is one of only two major bodies that can be shown to intrude strata below the Maentwrog Formation. It is bounded on the west by the Craiglaseithin Fault. The base of the intrusion rises through stratigraphically higher formations from south to north. Sedimentary rocks of the Clogau Formation dip under the intrusion in the north-west on the east side of the fault, but on the south and south-east there is a conformable upper contact between the intrusion and the Gamlan Formation. The doming of the superincumbent strata is apparent for up to 1 km to the south. In the north the upper part of the intrusion has penetrated through the Clogau Formation into the lower part of the Maentwrog Formation, though where not obscured by drift the bedding is subparallel to the adjacent contact. The Craiglaseithin Fault is presumed to post-date the intrusion, but smaller north-west faults in the area predate it; this may explain some of the relationships between the intrusion and the country rock in the northern parts. No contacts have been observed, though spotting (E 49499) is present [7387 3218; 7403 3307] close to the contact. The intrusion contains numerous barren quartz veins adjacent to its western margin, and veins containing sulphide minerals elsewhere have been tried for gold.

Nant Hîr

Porphyritic microtonalite is exposed beneath thick drift in the lower part of Nant Hîr and adjacent streams [752 318]. The form of the intrusion is unknown, though it appears that only the roof may be exposed in the core of an anticline. Contacts with the overlying Maentwrog Formation are exposed in the stream beds, and thermal metamorphic effects are discernible for about 1 m only.

Hafod Fraith

The intrusion of microtonalite in the Ffestiniog Flags Formation on the west side of the Afon Mawddach near Hafod Fraith [7484 2780] can be traced along strike for over 1 km. The upper contact is straight and parallel to the strike, whereas the base is irregular. The intrusion is 400 m thick near its north-eastern end where it terminates abruptly. Its outcrop tapers south-westwards where it is cut off by a fault.

In most of the intrusion the rock is a grey, medium-grained, sparsely porphyritic microtonalite, which in places is pyritic. Alteration is locally intense and is variable in nature. Sericitisation in association with the locally abundant development of chlorite, some after biotite (E 43776), is usual. In addition epidote in association with altered opaque minerals is locally abundant (E 43774, 49084).

A fine-grained, highly altered porphyritic phase of this rock

(E 43775) occurs along the base in the area north of Hafod Fraith, and at the top in the north-eastern part where quartz phenocrysts are abundant. South of Hafod Fraith malachite has been observed in the rock.

Coed Ty-Cerrig

In Coed Ty-Cerrig [723 248], between the Afon Gamlan and the main Dolgellau-Trawsfynydd road, an oval area of microtonalite represents a conformable lensoid intrusion within the lowest part of the Gamlan Formation. A small fault-bounded body of a similar rock exposed in Afon Gamlan at Rhaeadr-du appears to be at a slightly lower stratigraphic level but may be connected at depth. The northern body is 800 m long and 160 m thick in the middle.

The rock, which is unusually quartz rich, is texturally quite distinctive (E 49445, p. 49). PMA

Afon Wen

On the west side of Afon Wen (Figure 16), from its junction with the Mawddach northwards to Cwmhesian Isaf [7398 2696], there is an intrusion complex, mainly of microtonalite and porphyritic microtonalite, notable because of its association with disseminated copper mineralisation. For most of its length the complex, which reaches a maximum thickness of 500 m, lies concordantly within the Ffestiniog Flags Formation. It is no more than 180 m above the base of the formation south of Moel Dôl-Frwynog [742 249], but in the order of 300 m above it in the north around Foel Wen [746 266]. Here the intrusion is possibly continuous across a fault with the Hafod Fraith intrusion (p. 54). Between Moel Dôl-frwynog and Foel Wen there appears to be a discordant downward extension from the main body through the Maentwrog into the Clogau Formation. The area where the intrusion becomes discordant is covered with thick drift, but it has been drilled by Riofinex Exploration Ltd (Rice and Sharp, 1976) who confirmed the continuity of the intrusion complex. The mineralised body that they defined lies over this discordant downward extension.

In surface outcrops there is evidence of four rock types within the complex. The main one is non-porphyritic or sparsely porphyritic, medium-grained microtonalite. As at Hafod Fraith, however, a fine-grained porphyritic phase occurs locally along the base, eg. on Bryn Coch [738 245] and on Foel Wen. At both localities the rock is intensely altered. At the southern end of the body in Coed Bryn-prydydd [739 228] a separate sheet of porphyritic microtonalite (E 47480) lying above the main complex is apparently linked to it. This intrusion, which is cut by an intrusion of quartz microdiorite, distinctive in the effects of cataclasis within it, is spatially associated with several other thin sheets of the same rock exposed south of Coed Bryn-prydydd in the Afon Lâs. Around Bryn Coch small xenoliths of chloritic microdiorite are present in both the porphyritic and medium-grained microtonalite.

While drilling to investigate the copper deposit, Rice and Sharp (1976) identified three distinct phases of intrusion: an extensive 'older diorite' intersected by a 'porphyritic diorite', both cut by a relatively minor WNW-trending dyke called the 'Afon Wen diorite'. They also recorded intrusive breccias within the complex, and extensive brecciation of country rock and its assimilation into intrusive material in the roof.

Petrographically the 'Afon Wen diorite', though highly altered to carbonate, has much in common with the high-titanium dolerites described from other parts of the district (p. 50). The 'porphyritic diorites' compare exactly with porphyritic microtonalite. The 'older diorites', however, comprise a heterogeneous group of rocks mostly

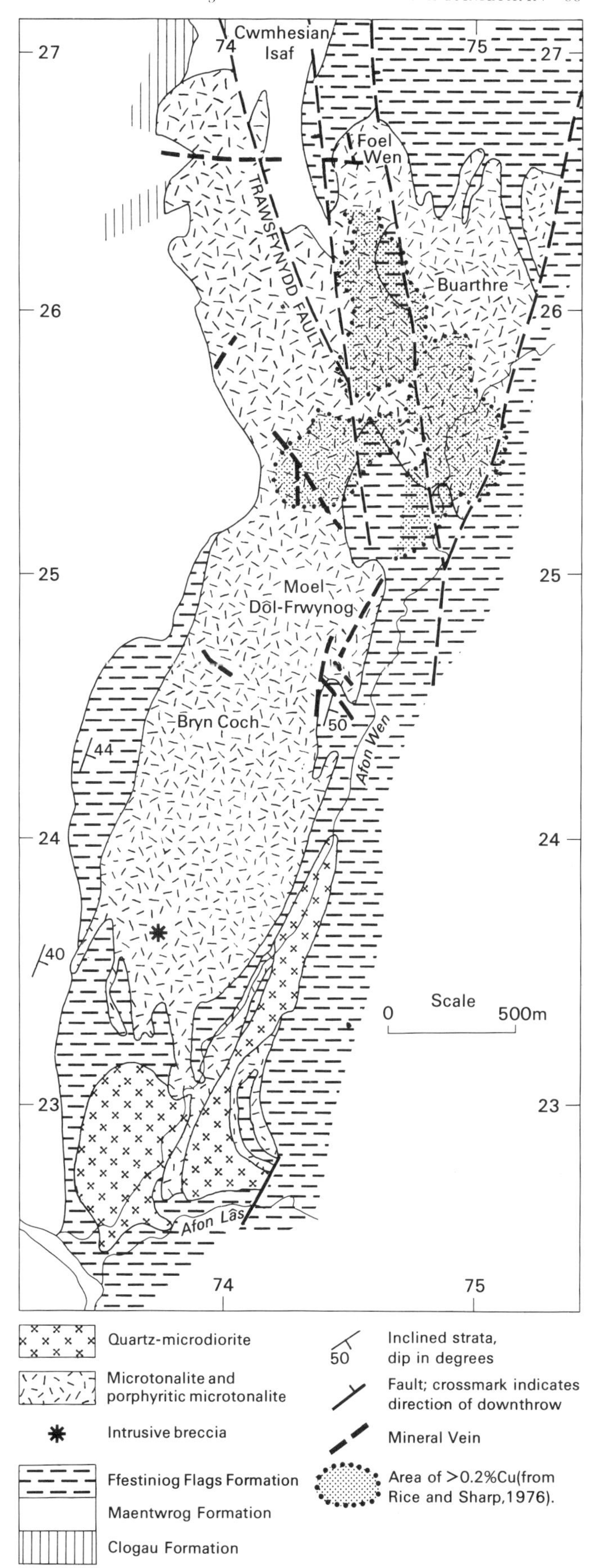

Figure 16 Sketch map of the Afon Wen intrusive complex

comparing with non-porphyritic microtonalite and quartz-microdiorite. Rice and Sharp (1976) recorded potash feldspar in these rocks and rare biotite. They also claimed that porphyritic varieties of the rock are present in the western part of the area they examined.

Hydrothermal alteration is more extensive in this intrusion complex than in any other examined except Hafod-y-fedw. There is extensive quartz veining in contact areas, and quartz-calcite veins are abundant within the orebody. Here, also, the rock contains fine hair-like fractures filled with chlorite and other gangue minerals in addition to pyrite and chalcopyrite (p. 81). The alteration conforms to the propylitic and phyllic types commonly recorded in association with porphyry copper mineralisation. Potassic alteration has not been proved, but the potash feldspar noted by Rice and Sharp (1976) may be indicative of this because it is unknown elsewhere in this suite of intrusions. Chlorite pseudomorphs after biotite are also present and locally abundant in the mineralised area, the discordant zone below it (E 49067), and elsewhere (E 48502, 48505). Some may be derived from primary biotite, which is uncommon in this suite of intrusive rocks; where biotite occurs in veinlets and irregular patches associated with sulphides (E 48505 from Bryn Coch; E 49067 from the discordant body) it is possibly secondary. Clay-rich parts of the deposit contain illite, chlorite and hydrobiotite (not kaolinite), and are thought to be due to later alteration associated with faulting and fracturing rather than the hydrothermal alteration associated with mineralisation (Rice and Sharp, 1976).

The pattern of alteration in the complex and its association with copper mineralisation is not fully understood. According to Rice and Sharp (1976) the mineralised body is within the zone of phyllic alteration though there is an irregular pattern to the alteration zoning. The widespread occurrence of secondary biotite, the presence of up to 0.2 per cent Cu in relatively unaltered rock at Bryn Coch, and lower levels of Cu in the intensely fractured and recrystallised quartz microdiorite (E 48502, 44746, 48499–501) in the lower Afon Wen suggest, however, that there may be more than one centre of mineralisation and alteration within this complex. PMA, WJR

Hafod-y-fedw

This intrusion complex to the south of the Tyn-y-Groes Hotel is entirely within the steeply eastward-dipping Maentwrog Formation. It is a generally concordant, faulted, sheet-like intrusion complex with a distinct upper component lying above a large enclosure of sedimentary rocks. In total it is about 500 m thick. It is xenolithic along the base [eg. 725 222] and contact rocks are baked. Quartz veins are abundant, especially south of Hafod-y-Fedw [7272 2189]. Some contain sulphides, and the Cae Mawr gold mine [725 225] worked one of them. Gilbey (1969) argued that the mineralisation was genetically unrelated to the vein mineralisation elsewhere in the Dolgellau Gold-belt (p. 84).

The two main primary rock types, quartz-microdiorite and porphyritic microdiorite, are variably and extensively altered. As in the southern part of the Afon Wen complex the rocks are commonly sheared. In one quartz microdiorite specimen (E 48495) both the fabric of the rock and individual plagioclase crystals are broken and penetrated by pale green chlorite.

The porphyritic microdiorite is always intensely altered. Phenocrysts of plagioclase are either recrystallised to a patchwork of albite or pseudomorphed either by muscovite or epidote. All the rocks show alteration to epidote group minerals, and in a sample from Cae Mawr Mine (E 45910) the rock is replaced by clinozoisite with some pale green chlorite and albite retaining only relict primary textures. A slightly less altered rock (E 48496) contains remnants of golden-brown hornblende phenocrysts, patchy calcite and chlorite/sphene pseudomorphs after biotite. In contrast to this essentially propylitic alteration, phyllic alteration (E 48498) has been recorded in a sill of porphyritic microdiorite within the sedimentary enclave near the top.

Dôl-fawr

The complex of quartz-microdiorite and microtonalite exposed on Foel Fawr intrudes the Ffestiniog Flags Formation as a series of sheets, in total about 450 m thick. The complex dips at moderate angles east-south-east and is separated by a fault from structurally complicated steeply dipping rocks to the north-east on Foel Cynwch. The lower part of the succession on the west side of Foel Fawr is concealed beneath scree. The upper contact can be traced along the side of a narrow valley [7324 2017]. The complex is pierced by one breccia pipe, and is separated from another by a fault.

In places the rocks in the complex are cleaved. Alteration is locally extreme and in one sample (E 44752) no primary constituents are recognisable.

The lowest exposed component in the complex is a sheet of medium-grained, sparsely porphyritic microtonalite above which a thin slice of sedimentary rock separates part of it from the main body of quartz microdiorite. A small body of fine-grained porphyritic microtonalite (E 48301), possibly a xenolith, occurs within the quartz-microdiorite [7308 2053]. Small xenoliths of similar rock occur nearby. Xenoliths of quartz microdiorite (E 44750) are present in microtonalite. The order of emplacement is likely to be, first porphyritic microtonalite, then quartz-microdiorite with, lastly, microtonalite. PMA

Subvolcanic intrusion complexes

Extending between Cae Poeth [760 257] and Berth-lŵyd [741 191] the western limit of the crop of the Rhobell Volcanic Group is bounded by intrusive rocks from the subvolcanic feeder zone to the Rhobell Fawr volcano. For descriptive purposes these rocks have been attributed to the Moel y Llan, Cerniau (Figure 17) and Nannau complexes, the latter probably the southern continuation of the Moel y Llan complex. They have been divided according to the nature of the country rock.

The main rock types in the complexes are porphyritic dolerite, rare amphibole-bearing dolerite, microdiorite, quartz-microdiorite and microtonalite. There are porphyritic and non-porphyritic varieties of the intermediate rocks. Some rocks are pyritic, containing cubes of 1 cm or more; others contain a little chalcopyrite. Age relations are consistent: on Moel y Llan and Cerniau xenoliths of dolerite occur in intrusions of all rock types, and intrusions of microdiorite and microtonalite intersect dolerite. In the Nannau complex, however, there are rare examples of dykes of highly altered dolerite cross-cutting all other rock types. All rocks are locally cleaved.

The emplacement of the Cerniau complex involved east–west dilation in excess of 1 km. The chilled boundaries of intrusions and the small proportion of xenoliths in them and in the lavas eliminate the possibility of loss of country rock either by assimilation or upward removal and ejection. In both Moel y Llan and Nannau, where brecciation of country rock in places may have preceded intrusion, the emplacement was not forceful and it is likely that intrusion took place in an environment of crustal tension. Uplift accompanied folding immediately prior to the eruption of the Rhobell Volcanic Group. Folding alone, however, cannot

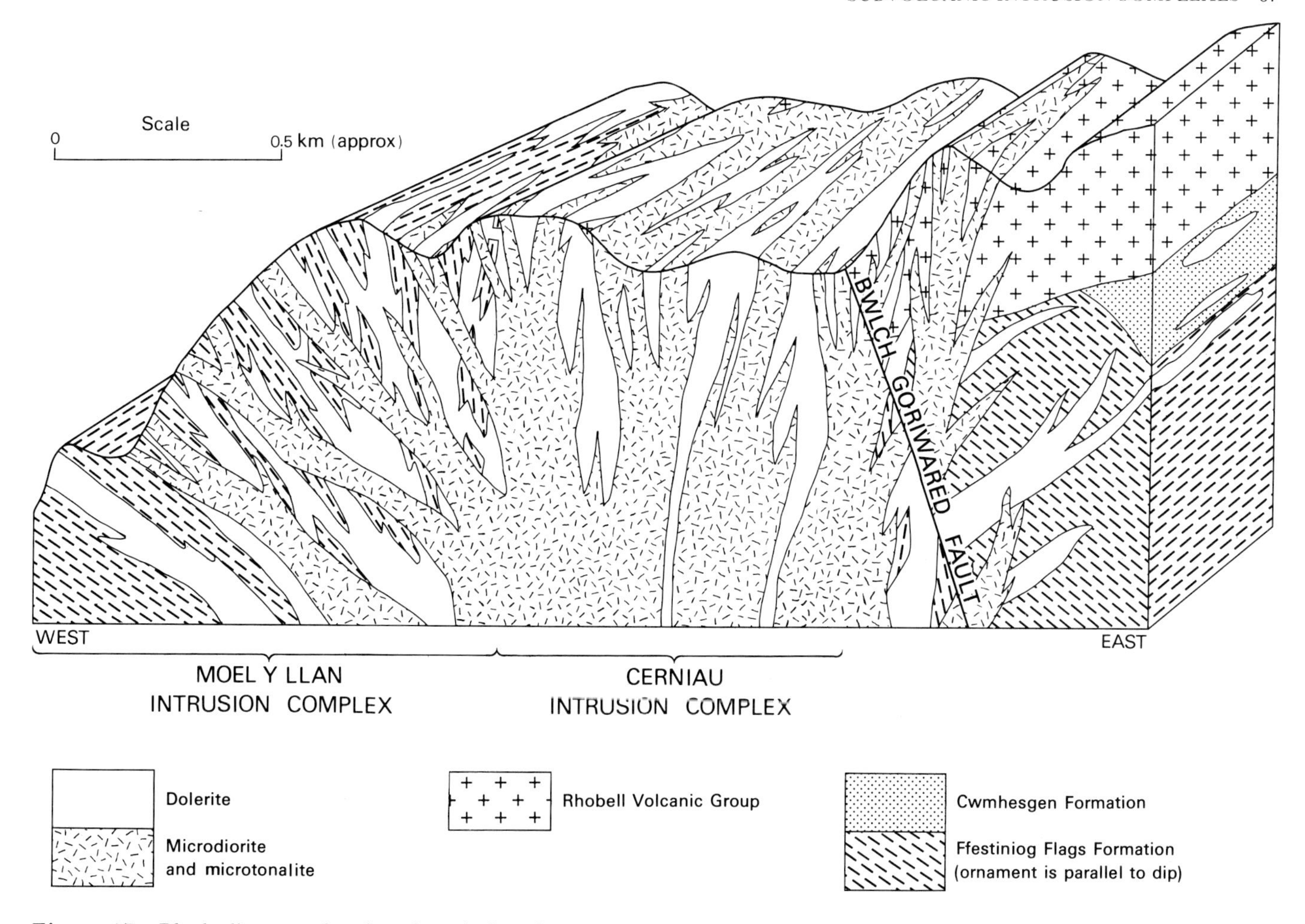

Figure 17 Block diagram showing the relations between Moel y Llan and Cerniau complexes, Rhobell Volcanic Group and the subvolcanic basement

have been responsible for the variable direction and amount of dip in the Ffestiniog Flags Formation caught up in the Moel y Llan and Nannau complexes. Apart from strata between the Wnion and Bala faults, vertical and overturned bedding is present only in these complexes. Immediately to the west of them dips conform to the regional pattern. It is suggested, therefore, that independent slices of country rock caught up between sheet-like intrusions were rotated during the process of emplacement. The controls on the orientation of the intrusions appear from the general concordance of the intrusions to have been bedding and possibly strike faulting, though the evidence for the latter has been obliterated by the intrusions. BPK, PMA

Autobrecciation and penecontemporaneous alteration affects all intrusive rock types in the Cerniau and Moel y Llan complexes. The breccias are generally restricted to poorly defined planar zones, parallel to the margins of the intrusions, up to 50 m in length and 10 m in width. The intensity of autobrecciation and subsequent alteration is variable, but it is not uncommon for the transition from unaltered brecciated rock to totally altered rock to take place across a zone less than 1 m wide. Within such zones jig-saw (crackle) breccia, in which fragments are separated by reticulate hairline fractures, are common. They are probably caused by late-consolidation retrograde boiling (Phillips, 1972). In places the passage of volatiles through the fractures has altered the fragment margins so that rounded, relatively fresh rock fragments are supported in a matrix of decayed rock. Continued streaming of volatiles has locally caused complete alteration of the original rock so that the brecciated appearance is lost. The uniform composition of the breccia fragments, the absence of tuffisitic intrusions, and the lack of sorting suggest that volatile streaming was not of sufficient vigour to produce a gas-fluidised system. These rocks were referred to by Wells (1925, p. 575) as 'crush-breccias'. The breccia zones may have accommodated some penecontemporaneous volcanic-tectonic strain, but the presence in some intrusions of unbrecciated cognate xenoliths within a brecciated host [7631 2411] suggests that tectonism played little part in the brecciation process. BPK

Brecciation of the wall rock in the Moel y Llan and Nannau complexes is not uncommon and in places the breccia fragments are incorporated in the intrusion. In parts of the Nannau complex the brecciation is demonstrably younger than the intrusions, and there is evidence that the breccias themselves are intrusive (p. 59). PMA

Cerniau

The Cerniau intrusion complex is bounded on the east by the Bwlch Goriwared Fault (Figure 17) and on the west by the easternmost occurrence of sedimentary country rock. The latter boundary forms a regular and continuous surface dipping between 75°E and vertical. On the east of the Bwlch Goriwared Fault there are numerous intrusions, mostly of microtonalite, within the Rhobell Volcanic Group lavas; these are thought to be a high-level expression of the Cerniau complex. Within the complex small pockets of locally bedded lava breccia occur in the central and topographically highest parts [7630 2411], but elsewhere there is no country rock. The component intrusions are elongate, lenticular in cross-section, and strike a little east of north. Their steeply inclined attitude can be determined by the dip of flow-banding, the flow-alignment of phenocrysts and cognate xenoliths, and the few exposed contacts. Contacts are sharp and there is usually no rheomorphism or brecciation at them.

Moel y Llan

This complex consists of numerous concordant or semi-concordant intrusions alternating with Ffestiniog Flags Formation. The intrusions dip 25° to 40° east in the west, and steepen towards the boundary with the Cerniau complex becoming vertical or overturned. The strike, however, remains constant even among thin slabs of sedimentary rock entrapped between intrusions. There is an eastward increase in the proportion of intrusive rock and in the amount of disruption across the complex. Xenoliths of sedimentary rock are commonly crowded in a narrow zone at the margins of the intrusions. Contacts are commonly sharp and thermal metamorphism limited to a zone no more than a few centimetres wide, but in places the adjacent wall rock is brecciated along the length of intrusions and locally along strike beyond their terminations. These breccias are commonly devoid of intrusive material and some are zoned inwards from slightly displaced blocks to a jumble of angular fragments. In places the breccia is incorporated in the intrusion which is not itself brecciated. BPK

Nannau

As with Moel y Llan, entrapped slabs of Ffestiniog Flags Formation occur throughout this poorly exposed complex, and it is probable that the two complexes are physically connected beneath the drift-covered ground south-west of Llanfachreth. There are many comparisons between them: intrusions are concordant or semi-concordant, the proportion of sedimentary country rock diminishes eastwards, bedding strike is consistent but dips vary. On Foel Cynwch bedding dips from 27° to steeply east and west, and is locally overturned. Though there are minor open folds in the sedimentary rocks the pattern of bedding orientations is not readily interpretable in terms of folding. On Foel Cynwch the intrusions are sheet-like or thin lenticular, commonly between 0.3 to 2 m thick. In places slices of sedimentary rock as little as 1 m thick are trapped between them [7388 2162]. Farther east, where enclosed country rock is uncommon, the size of intrusions is not possible to determine. The close association between intrusion and breccia observed on Moel y Llan is repeated here; e.g. wall rock is brecciated and veined by intrusion at the northern end of Precipice Walk [7394 2168], and elsewhere the margins of intrusions are xenolithic [7337 1967]. PMA

Minor concordant intrusions within the Cambrian strata

The intrusions vary in thickness from a few centimetres to over 50 m, but a large proportion are less than 5 m thick. They are generally concordant, although boundaries are locally transgressive on a small scale. Rarely, a sill may pass laterally into a dyke or there may be discordant apophyses adjacent to a sill. Most of the intrusions are tabular and laterally persistent, though thin, lensoid forms occur. Terminations vary from tapering to abrupt (Plate 2.2). Some thick sills show complex interfingering with the country rock which may be a result of multiple injection. There is evidence of intrusion into wet or unlithified sediment at stratigraphic levels as low as the upper part of the Maentwrog Formation: in Nant Ganol, for example [7585 3263], there are contorted margins to thin sills, with flames, disconnected blobs of igneous rock and contorted, folded sedimentary wall-rock. Xenoliths are uncommon and with rare exceptions consist of wall rock or cognate xenoliths. A xenolith of coarse-grained, leucocratic, two-feldspar granite was recovered (E 40854) from a sill east of Trawsfynydd. Both the chilled margin and the thermally altered wall-rock zone rarely exceed 1 cm thick.

All the sills are pre-cleavage, though usually only the margins are cleaved. In several places they are cut by NW-trending dolerite dykes; e.g. near Clogau goldmine [6752 2049].

Petrographically and compositionally there is little to distinguish the dolerite and the intermediate rocks in these sills from the dykes which comprise the high level Moel y Llan, Cerniau and Nannau intrusion complexes (p. 50). PMA, WJR

There are numerous dolerite intrusions within the sedimentary rocks immediately below and in contact with the base (or the projected plane of the base) of the Rhobell Volcanic Group. These are concordant and semi-concordant with respect to the base of the volcanic group and, therefore, may be termed sills. In places these sills transgress into the volcanic group [7830 2204, 7790 2868]. They are commonly intensely cleaved and highly susceptible to weathering. In cross-section they are irregular, roughly elliptical or lens shaped, and they display a broken form which is due to their cross-cutting relationships with the sedimentary host rocks.

These intrusions are present on the north, east and south of the main outcrop of Rhobell Volcanic Group, and may originally have formed a continuous network of intrusions. They are clearly related to that magmatic event and constitute a distinctive very high-level sill regime, to which the Bryn Brâs laccolith (p. 53) also probably belongs. BPK

Dykes

Dykes of dolerite are most numerous in the area between Llyn Eiddew-mawr [646 337] and Llyn Trawsfynydd, and on Craig Ganllwyd [708 258] north of Afon Gamlan where they occur in swarms with a dominantly north-westerly trend, cutting rocks of the Harlech Grits Group. These two areas lie on either side of the Dolwen pericline, but there is no continuity of the swarms across it. Dykes with this same trend cut younger Cambrian rocks, especially in the north-east of the district, but they are uncommon. According to Matley and Wilson (1946) they are usually less than 7 m wide, although some reach 14 m, and they cannot normally be traced for more than 2 km. A complementary set of north-east dykes, which is apparently older, occurs in both the main areas of intrusion. The north-easterly dykes are extremely altered. The NW-trending dykes are relatively fresh and show petrographic characteristics of two main compositional types. One type resembles the ophitic ilmenite-rich augite-dolerite among the sills, the other is ilmenite-free, though these rocks are more altered. WJR, PMA

Dolerite dykes occur throughout the Rhobell Volcanic Group, though they are difficult to distinguish except where they cut breccias. They have chilled margins 1 to 2 cm wide. Autobrecciation and penecontemporaneous alteration occur in places [7817 2645]. Dykes tend to branch where they cross breccias [7860 2606, 7810 2555]. Elsewhere their outcrops are linear or lenticular [793 266], and there is no preferred orientation. Xenoliths are uncommon, but include coarse-grained greywacke [7855 2603] derived from the Harlech Grits Group and heterolithic breccia derived from the underlying volcanic group [7826 2648]. BPK

Dykes in the Aran Volcanic Group are uncommon. Excluding discordant basaltic intrusions (E 42173) which are spatially associated with extruded pillow basalts, there is a dyke in the Blaen Lliw area [8014 3369] of sparsely porphyritic, intensely altered dolerite (E 47470) petrographically similar to the sills within the group. In the same area [8019 3355], a thin dyke of feldspar porphyry (E 43139) intrudes the Benglog Volcanic Formation. PMA

Intrusions in the Aran Volcanic Group

Within the district there are many intrusions of ophitic augite dolerite in and just below the Ordovician succession. Other intrusive rock types are rare; they include a small body of microtonalite on the southern side of Ffridd yr Allt-lŵyd [801 290] and the edge of a large intrusion of microdiorite in the upper part of Nant Helygog [802 183].

The dolerite intrusions are present throughout the Aran Volcanic Group, but are absent from strata above it, and they occur at and just below the basal Arenig unconformity. Most of the intrusions lie within mudstone formations.

Though locally transgressive, the overall form of the intrusions is sill-like. They vary laterally in thickness and locally follow faults. The thickest and most extensive intrusions occur at the basal unconformity. They are coarse-grained, locally gabbroic, and possess well-developed chilled margins. Leucocratic pegmatitic schlieren up to 1 m thick and veinlets are not uncommon in them; e.g. in Afon Wnion [7865 2037]. In the upper parts of these intrusions there are melanocratic pegmatitic schlieren and veins; e.g. in the large body near Brithdir [7826 1940, 7704 1792] and in a smaller intrusion to the east [8020 1909].

There is a general decrease in size of intrusion with stratigraphic height. The higher intrusions are fine-grained; vesicularity increases upwards, and the rocks are more altered than the large, lower intrusions. Dunkley (1978) provided evidence from the Aran Mountains, east of the district, that some stratigraphically high intrusions were emplaced within wet sediment.

Xenoliths are rare. The intrusions show good columnar jointing in places (Plate 7.2). They are commonly altered at their margins, but thermal effects in country rock are minimal. Spotting has been observed in Nant Helygog [7868 2012] and in the Afon Melau [7879 2389] . The intrusions are folded and cleaved at the margins.

The microtonalite on Ffridd yr Allt-lŵyd [801 290] appears to be unique within the district. It is a small oval-shaped plug intruded into the volcanic sandstones of the Allt Lŵyd Formation. The rock consists of phenocrysts of albite/oligoclase in a devitrified glassy groundmass consisting of quartz, feldspar and chlorite. Vesicles are filled with an outer rim of quartz and a core of chlorite or calcite. Clasts of this rock occur as xenoliths, some up to 30 cm long, in the tuffs of thc Benglog Volcanic Formation. This rock is similar to the quartz latites of the Arenig and Migneint areas to the north (Lynas, 1973; Zalasiewicz, 1981), where it forms large intrusions in the lower part of the Ordovician succession within and below the level of the Serw Formation (equivalent to the Benglog Volcanic Formation). PND, AAJ

Intrusive breccias

Six bodies interpreted as pipes occur in the southernmost part of the belt; two on Foel Fawr [7290 2038, 7290 2028], two on Foel Cynwch [7335 2098, 7350 2089], one on Mynydd Penrhos (Plate 11.1) [7376 2366], and the sixth at Glasdir [741 224] (Plate 11.2). They are oval or irregular in cross-section, ranging in size from 60 to 200 m in longest diameter and, except for the one at Glasdir, they are unproved in size and shape at depth. They have sharp intrusive contacts with the country rock, which in two cases is a major acid or intermediate intrusion and in the others is sedimentary rock of the Ffestiniog Flags Formation. Three of them form prominent rocky knolls. The Glasdir body consists only of angular fragments of the adjacent country rock; the spaces between are filled with secondary minerals. The other five pipes, however, contain rounded exotic fragments in addition to fragments of wall rock and there is evidence of an original clastic fine-grained matrix.

The greatest concentration of dyke and sill-like breccia bodies, which are rarely more than 1 m and may be as small as 5 cm wide, is within the Nannau complex (Figure 5). In many instances the breccias occur along the top or bottom margins of sills of intermediate and, in one case, basic intrusive rocks and some can even be traced beyond the termination of a sill; e.g. on Bwlch-y-Bi [7776 3522]. Just south of Moel y Feidiog, and in particular on Foel Cynwch, several dykes occur within the Ffestiniog Flags Formation trending parallel to the strike; elsewhere they occur within intrusions. In many places the form of the intrusive breccia is not known.

The majority of these intrusive breccias contain only angular and subangular fragments of rock types similar to the wall rocks (Plate 10.1), but there is usually evidence of movement of the fragments. For example, in intrusive breccias along the margins of igneous sills it is possible to find two outer bands, one adjacent to the sedimentary wall-rock and composed of fragmented unrotated sedimentary rock, the other adjacent to the igneous rock and composed only of fragments of that rock. Between them, and showing sharp contacts, is a band of mixed sedimentary and igneous fragments, clearly intruded into place. In a body of this type in the upper reaches of Afon Gain [7660 3487], at the margin of the Ranges microtonalite intrusion, the thermal metamorphic spotting in the aureole to the Ranges intrusion clearly

Plate 11
1 Intrusive breccia from a pipe on Mynydd Penrhos containing igneous and sedimentary fragments set in a largely clastic matrix. (Neg. No. 11543).
2 Intrusive breccia from Glasdir copper mine with a matrix of chlorite mainly, and some highly reflective sulphide (Neg. No. 11545).

predates the brecciation. There is no clastic material in the matrix (E 41979), which is composed entirely of chlorite and quartz with pyrite, similar to the Glasdir breccia.

In some intrusions the fragmental material is different from the adjacent country rock, but its composition and angularity suggest that it has not moved far from source; e.g. in the south of the district [7326 1834] a dyke of angular sedimentary fragments cuts porphyritic microdiorite.

In three of the intrusive-breccia sills that were examined various proportions of rounded exotic fragments have been identified in addition to the angular wall rock fragments. Those on Bwlch-y-Bi [7777 3506] and near Cynwch [7392 2074] occur along the contact of fine-grained porphyritic microdiorite; a third overlies a large body of dolerite on Moel-y-Feidiog [7850 3148]. In all three the exotic material, as in the pipe-like bodies, is composed of intrusive rocks common in the area in general, but not present at surface in the immediate vicinity of the intrusive breccia. The matrix is recrystallised quartz, chlorite, sericite and feldspar with some remaining angular grains of quartz and feldspar suggesting an original clastic origin.

The origin of intrusive breccias has been subject to some controversy over the years and the literature on it is extensive. Allen and Easterbrook (1978) briefly review it. They conclude that in this area the intrusive breccias were probably formed as a result of fumarolic activity during the waning stages of the Rhobell Fawr volcanism in the late Tremadoc.

Wolfe (1980) has recently pointed out that fluidisation has incorrectly been invoked to explain the mode of emplacement of intrusive breccias. In his view most pipes result from phreato-magmatic explosions when vast quantities of meteoric water come into contact with hot magma at depths of less than 4 km. The mode of intrusion of the fragments in the breccias is mainly ballistic. Such a mechanism might be applicable to the intrusive breccias in this district, though the role of late hydrothermal liquids, from which a wide variety of minerals crystallised including copper and iron sulphides in the interfragment spaces, must at some stage have been important, either mixed with meteoric water during the explosions or immediately afterwards.

DETAILS

The Glasdir body, which is mineralised at the margin (p. 82), has been described by Allen and Easterbrook (1978). It is oval in section measuring 200 × 100 m, and has been proved to 210 m depth in the old copper mine. It is composed of randomly oriented angular fragments, mostly less than 30 cm diameter, of rocks from the adjacent Ffestiniog Flags Formation and interstratified intrusions (Plate 11.2). The matrix is mainly composed of chlorite and quartz, in places crystallised concentrically between fragments as though they grew in cavities, with minor pyrite, muscovite and apatite.

In the other five pipe-like bodies, rounded exotic fragments of various kinds are mixed with randomly orientated, subangular blocks of adjacent sedimentary and igneous rocks. These reach 1.5 × 1.2 m in size in one of the Foel Cynwch bodies [7335 2098]. The exotic blocks usually consist of varieties of intrusive rock not found at surface in the immediate neighbourhood, though always within the compositional range of the late Tremadoc intrusive episode. In the body on Mynydd Penrhos [7376 2366] there are fragments of coarse-grained sandstone similar to rocks in the Harlech Grits Group.

The matrix in these pipes is usually fine-grained, recrystallised quartz, feldspar, chlorite and sericite, but evidence of an original clastic fabric is present in small angular grains of quartz and feldspar usually in the range of fine sand. Locally (E 48303) the matrix is highly feldspathic. In most rocks there are small, irregular-shaped patches of chlorite, some marginal to enclosed fragments (E 48300, 48503, 44754), or coarse aggregates of chlorite, quartz and lesser muscovite (E 48300, 48503) similar to the matrix of the Glasdir breccia. Epidote is uncommon (E 48503). The effects of hydrothermal alteration are evident in some rock fragments which are totally sericitised and others which are replaced by calcite (E 48298–9). Veins with calcite, quartz and chlorite, and albite are present in some fragments. The margins of fragments in some highly altered rocks are corroded and intergrown with the matrix (E 48299). Pyrite is ubiquitous. It is mostly confined to the matrix, commonly associated with chloritic patches, and in some rocks (E 48300) it is concentrated at the margin of rock fragments. Besides Glasdir, only one other pipe, on Foel Fawr [7290 2038], contains chalcopyrite.

A body of unknown form is exposed in a road cutting parallel to Afon Lâs [7415 2259] and in the stream below the road. The intrusive breccia, which is 6 m at its thickest point, intrudes quartz microdiorite. In its upper part it is composed entirely of fragments of Ffestiniog Flags Formation (E 49052), whereas in its lower part (E 46056–7) there are also fragments of igneous rocks and plentiful coarse-grained clastic quartz in the matrix. The rock is intensely hydrothermally altered; both pyrite and chalcopyrite are present, the latter mostly in veinlets.

PMA

CHAPTER 5

Structure

The conclusion of a number of authors, including Shackleton (1952) and George (1963) was that the British Caledonides are the product of essentially episodic deformation throughout the Lower Palaeozoic. This is supported in the Harlech Dome where the first of four fold-phases recognised took place in the late Tremadoc; the last is the major end-Silurian event. There is, in addition, a long history of faulting. A tentative chronology of structural events is shown in Table 9.

FOLDING

The western area

Matley and Wilson (1946) identified and named four major structural features in the area that they mapped (i.e. west of the Llanelltyd–Trawsfynydd road): the Dolwen Pericline; the Caerdeon Syncline; the Traeth Bach Syncline (Figure 18); and the Coastal Anticline. The Dolwen Pericline is not simple, but is made up of a number of parallel folds arranged en-echelon, giving the main fold axis a sinusoidal trace. The axis is traceable into the Ordovician cover as far north as Blaenau Ffestiniog and dies out northwards on the flanks of the W-trending Dolwyddelan Syncline. The Caerdeon Syncline is also made up of a number of smaller folds. The structure plunges southwards, and terminates to the north of the Moelfre Fault. North-west of the Upper Artro and Coed Gerddi faults it continues across Moel Goedog towards Talsarnau where it passes into the north-plunging Traeth Bach Syncline. The Cambrian outcrops are now known to be truncated against the Mochras Fault, east of Morfa Dyffryn, so that the 'Coastal Anticline' which Matley and Wilson (1946) proposed as a southward continuation of Fearnsides' (1910) Ynyscynhaiarn Anticline flanking the Caerdeon Syncline is cut out by the Mochras Fault; geophysical evidence (p. 69) suggests that the basement rocks form a syncline under the Mesozoic rocks of Tremadoc Bay.

The eastern area

East of the Trawsfynydd–Llanelltyd road, Wells (1925) identified the dominant structures as being a syncline through Rhobell Fawr and an anticline (Melau Anticline in Figure 19) in the upper Cambrian strata to the east of Rhobell Fawr. The present survey has shown that the first of these structures is a synclinorium, the axis of which can be defined by the outcrop pattern, though in detail individual folds are not traceable for very far. The axis of the associated anticlinorium, defined on outcrops, runs from Afon Melau to Rhiw Felen and northwards to Moel Feidiog. Wells (1925) claimed that the Rhobell Syncline might represent the collapse of the 'arch' which supported the erupted material. The apparent convergence of the axes of the main folds to the north of Rhobell Fawr may support this hypothesis, though both of the axes can be traced into the underlying Cambrian strata.

The structural style is controlled by the competent horizons. The formations of sandstone and volcanic rocks show broad open periclinal folds, in marked contrast to the more complex folds with short wave length in the incompetent beds. The Dolgellau Member, in particular, shows a variety of fold styles. The predominant form is the small scale, paired, asymmetrical anticline and syncline. The

Table 9 Correlation of fold phases described by various authors

East Harlech Dome	Migneint area Lynas, 1970	Llwyd Mawr Syncline Roberts, 1967	North Wales Shackleton, 1952
	F4 strain slip F3 not seen	F3 & S3 NW-trending F2 & S2 NE-trending	
4 NE-trending folds (End-Silurian)	F2	F1 & S1 NE-trending	Main Caledonian deformation (NE-trending)
3 NNW-trending folds (?late Ordovician - Silurian	F1 (NW-trending)		End-Ordovician movements (Taconic phase, NE-trending)
		?Bedding plane cleavage	Mid-Ordovician movements (Trondheim phase)
2 North-trending folds (Post-Rhobell Volcanic Group, Pre-Arenig)			Pre-Arenig movements (Tryssil phase, ? NE-trending)
1 NNE- to north trending folds (Pre-Rhobell Volcanic Group)			

steeper limb may be overturned. The folds are commonly associated with brittle fracture, usually in the axial region, and high angle reverse faulting. Monoclines and box folds are the other common fold forms. They generally have associated faults in the axial regions. On a regional scale the Ffestiniog Flags Formation behaves in a competent fashion. In detail, the coarse quartzose siltstones behave as competent beds within the grey cleaved siltstone and are characterised by boudinage. The lower part of the Maentwrog Formation behaves in a similar manner. Tectonic ripples, caused by the interference of cleavage with bedding planes on coarse quartzose siltstone beds, are common in this formation. AAJ

The structural evolution of the eastern part of the district has been established by detailed survey. Four phases of folding can be identified and are dealt with in turn below.

NNE to N-trending folds (Pre-Rhobell Volcanic Group)

Around Rhobell Fawr the volcanic rocks progressively overstep, to the west, older rocks within the Mawddach Group and there is evidence that the pre-Rhobell deformation involved folding. Near Hafodty-hendre [7655 2745] rocks of the Cwmhesgen Formation are preserved in the core of a NNE-trending syncline that is overstepped by the Rhobell Volcanic Group. Within the Moel y Llan intrusion complex and on Foel Cynwch, the western part of the Nannau intrusion complex, the bedding in the Ffestiniog Flags Formation is locally steeply dipping to overturned. Bedding attitudes are not all easily interpretable in terms of folding (p. 56). On Foel Cynwch there are, however, N-trending tight, slightly overturned folds with axes plunging up to

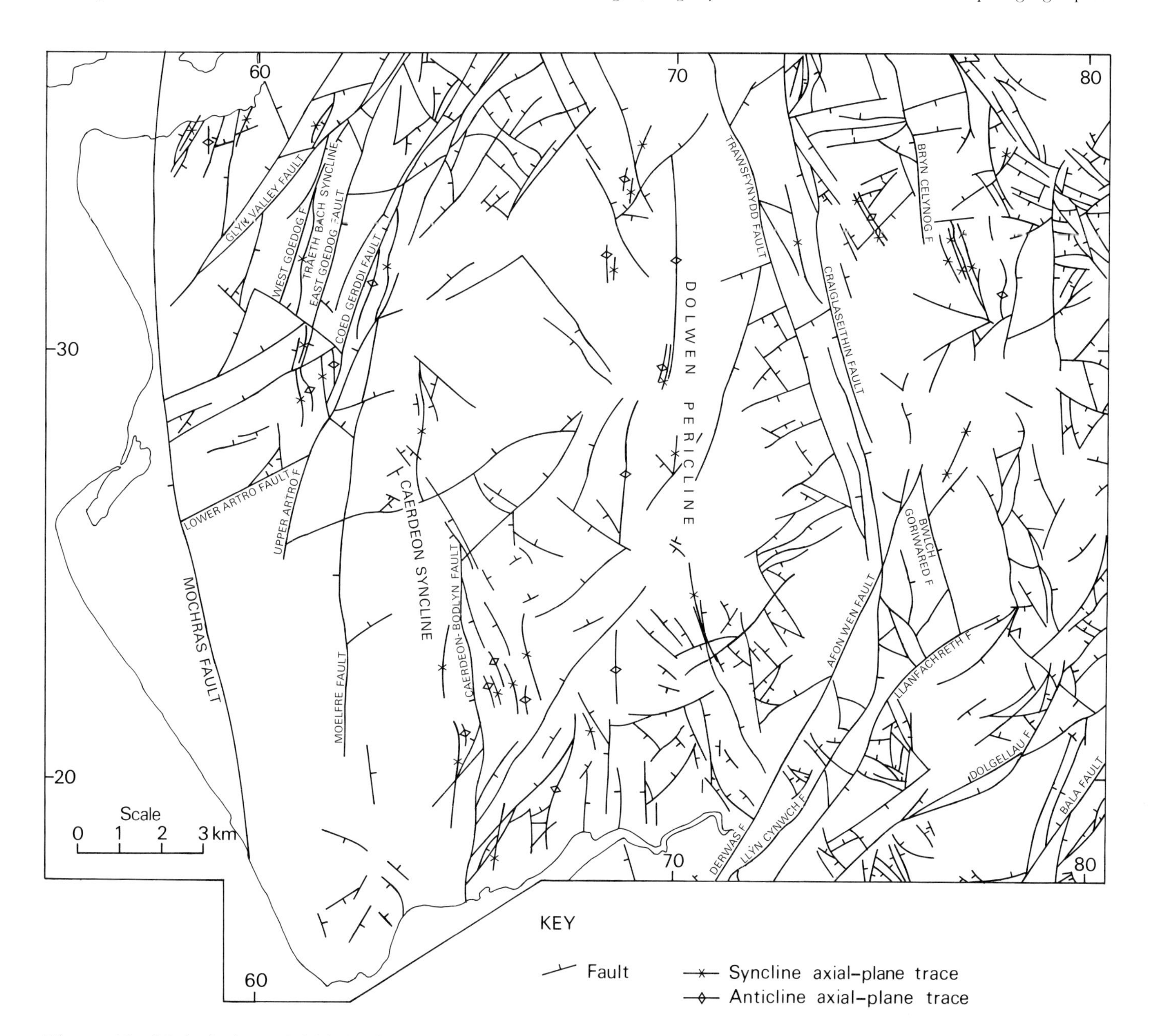

Figure 18 Main faults and folds in the Harlech area

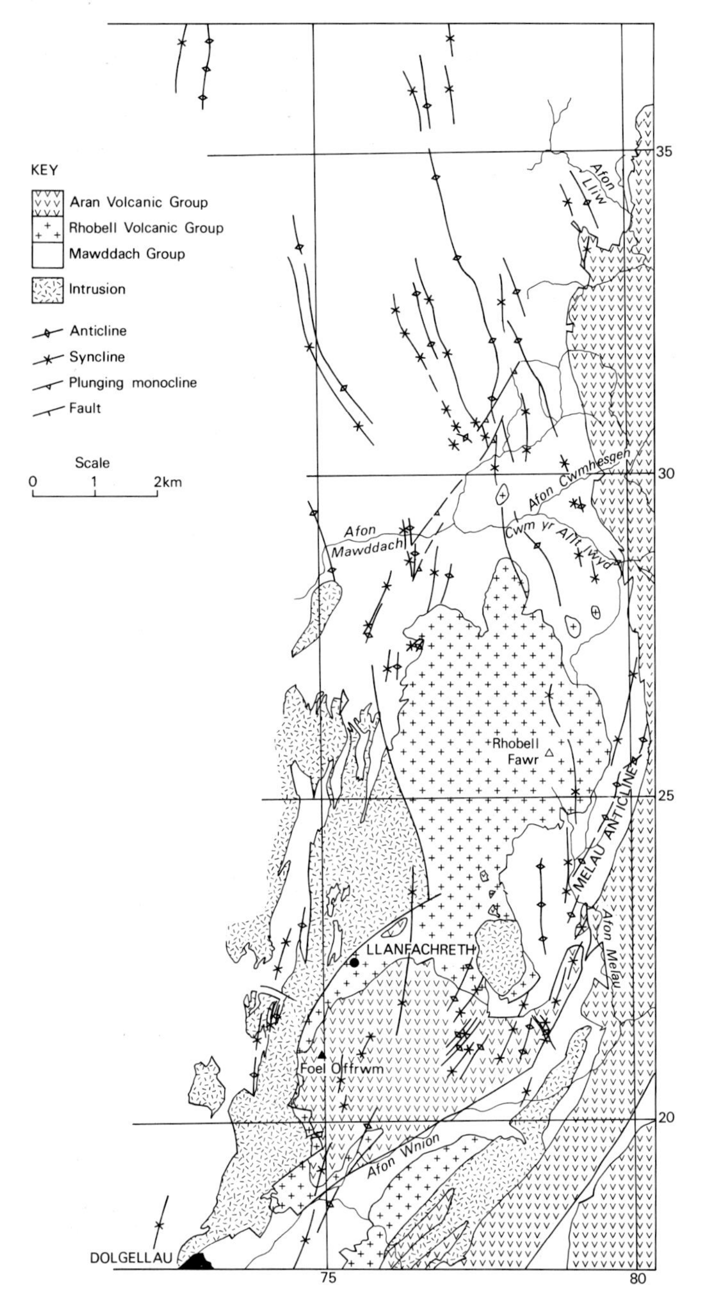

Figure 19 Axial plane traces of major folds in the upper Cambrian and lower Ordovician east of the Harlech Dome

28°S; adjacent outcrops show subparallel, open folds with near horizontal axes and an axial-plane cleavage. The overturned folds are interpreted as early, probably pre-Rhobell, folds modified during the deformation that produced the open folds.

Kokelaar (1977, 1979) suggested that the pre-volcanic deformation included the formation of an eastward-facing monocline in the Ffestiniog Flags Formation and tight folding of the overlying Cwmhesgen Formation, and that the folding developed in relation to movement of N-trending fault-bounded blocks during NW–SE compression. He includes among the N–S faults the Afon Wen Fault (trend 030°), which is of some importance in this area for steep dips and overturned bedding are restricted to a belt 1 to 2 km wide to the east of it. It is likely that deformation continued throughout the period of eruption, leading to the further rotation of fault blocks and the local up-ending of slabs of country rock during intrusion late in the eruptive sequence.

The pre-Rhobell Volcanic Group folding is the earliest for which there is good evidence and, though in intensity it is most marked around Rhobell Fawr, the deformation (involving block faulting, tilting and faulting and culminating in uplift and erosion in immediate pre-Arenig times) is regional in context. On St Tudwal's peninsula on the Lleyn the basal Arenig rocks overlie unconformably both the Harlech Grits and the Mawddach groups; Nicholas (1915) postulated a pre-Arenig fold phase to account for this unconformity.

Twenty kilometres to the north-east in the Deudraeth area, however, Fearnsides and Davies (1944) were unable to prove an unconformity between the uppermost rocks of Tremadoc age and the Garth Grit. An equal distance to the south-east, around Arthog, Cox and Wells (1921) concluded that, at the most, there was only slight discordance at this level. Both of these areas lie along projections of major synclinal axes; Caerdeon in the south and Traeth Bach in the north. In the vicinity of Penmaenpool, however, along the southward prolongation of the axis of the Dolwen Pericline, the Rhobell Volcanic Group and the lowermost Ordovician rocks overstep the Dolgellau Member. It is evident, therefore, that the Caerdeon–Traeth Bach axis and the Dolwen Pericline were both important structural lines active before the eruption of the Rhobell Volcanic Group.

N-trending folds (Post-Rhobell Volcanic Group, pre-Arenig)

Wells (1925) and Kokelaar (1977, 1979) examined the relationship between the outlier of Allt Lŵyd Formation (Basement Group) that lies unconformably on the Rhobell Volcanic Group at the summit of Rhobell Fawr; they concluded that the volcanic rocks had been folded into a N-trending syncline prior to the deposition of the Garth Grit Member. Wells (1925) considered that this deformation most probably related to the collapse of the underlying magma chamber when eruption ceased. Though the fold axes are traceable into basement rocks to the north some collapse must have taken place in order to lower the top of the volcano to sea level during the Arenig transgression.

It is difficult to determine the regional extent of this deformation and to distinguish it from pre-Rhobell folding in areas where the Rhobell Volcanic Group is not present. For example in the area between Afon Cwmhesgen and Llechwedd Rhudd [802 353] the Dol-cyn-afon Member varies between 135 to 150 m in thickness, but in the north-east corner of the district there is a sudden increase to 300 m, continuing eastwards to more than 500 m on the north flanks of Moel Llyfnant. In this area the base of the Aran Volcanic Group dips at about 20° less than the average for the bedding in the Dol-cyn-afon Member. About 10 km NNW in the Migneint, however, Lynas (1973) was able to demonstrate local conformity between the equivalent of the Dol-cyn-afon Member and the Garth Grit. He argued that sedimentation at the end of the Cambrian was strongly controlled by N–S block faulting, and that pre-Arenig folding was not much in evidence in the Migneint. PMA, AAJ

NW-trending folds (?late Ordovician to Silurian)

These folds, trending NW to NNW, are best developed within the Cambrian succession to the north of grid line 30, and are only present locally within the Ordovician to the south. The folds are asymmetrical with the steeper limb dipping eastwards. They have a wide axial angle and plunge to the south. There is an associated axial-plane cleavage. Within the Dolgellau Member [7927 2883] the earlier folds face downwards along this cleavage. Evidence for the age of these folds is scarce but they are crossed by NE-trending folds ('plunging monoclines') north of Rhobell Fawr which may be coeval with the similar trending main regional folds of this area.

NNE to NE-trending regional folds (End-Silurian)

These periclinal folds are best seen within the Ordovician cover. They are open folds with axes trending 020° to 030°,

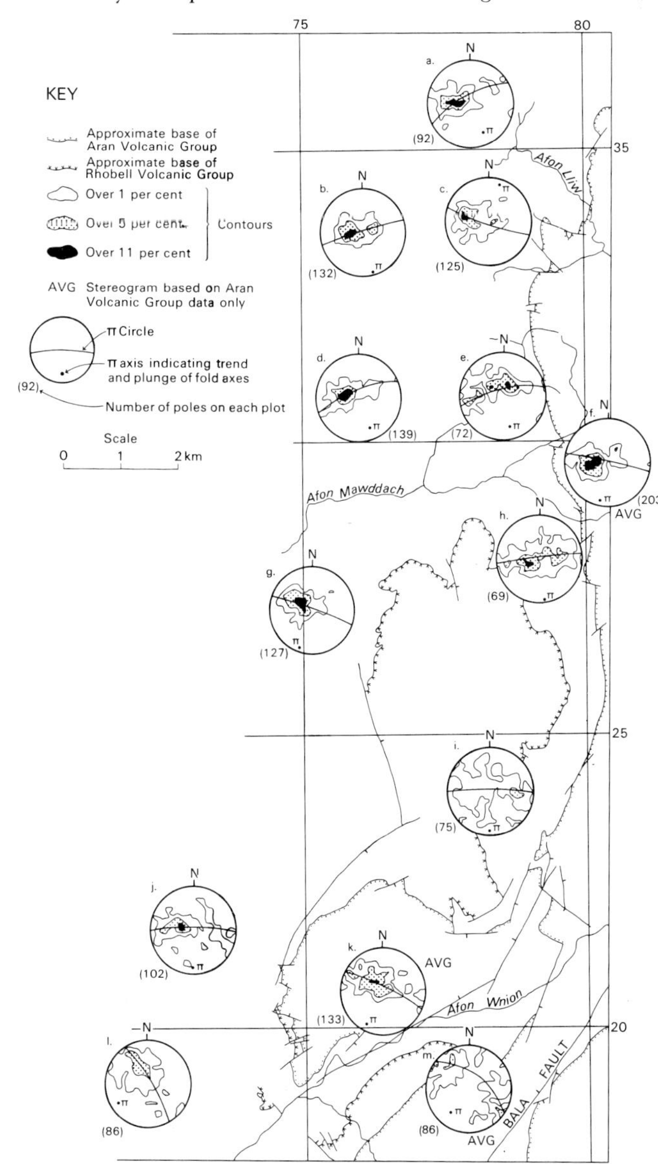

Figure 20 Contoured stereograms showing poles to bedding in the eastern Harlech Dome

and plunges vary between 20°NE to 40°SW. This deformation may have caused a tightening of the earlier structures, and a small circle spread on the πS_0 along a NE-trending axis can be seen in Figure 20. There is an associated NE-trending cleavage. The NW-trending cross-flexures or plunging monoclines within the Cambrian sedimentary rocks to the north of Rhobell Fawr may have been produced by this phase of deformation. The relationship between the earlier NW-trending folds and the NE-trending flexures can be demonstrated in a quarry [7485 3093] where NW-trending small-scale folds are refolded on an axis trending approximately 078° and plunging at 73° north-east (Figure 22).

The major structures of the district, the Caerdeon Syncline and the Dolwen Pericline, though initiated during

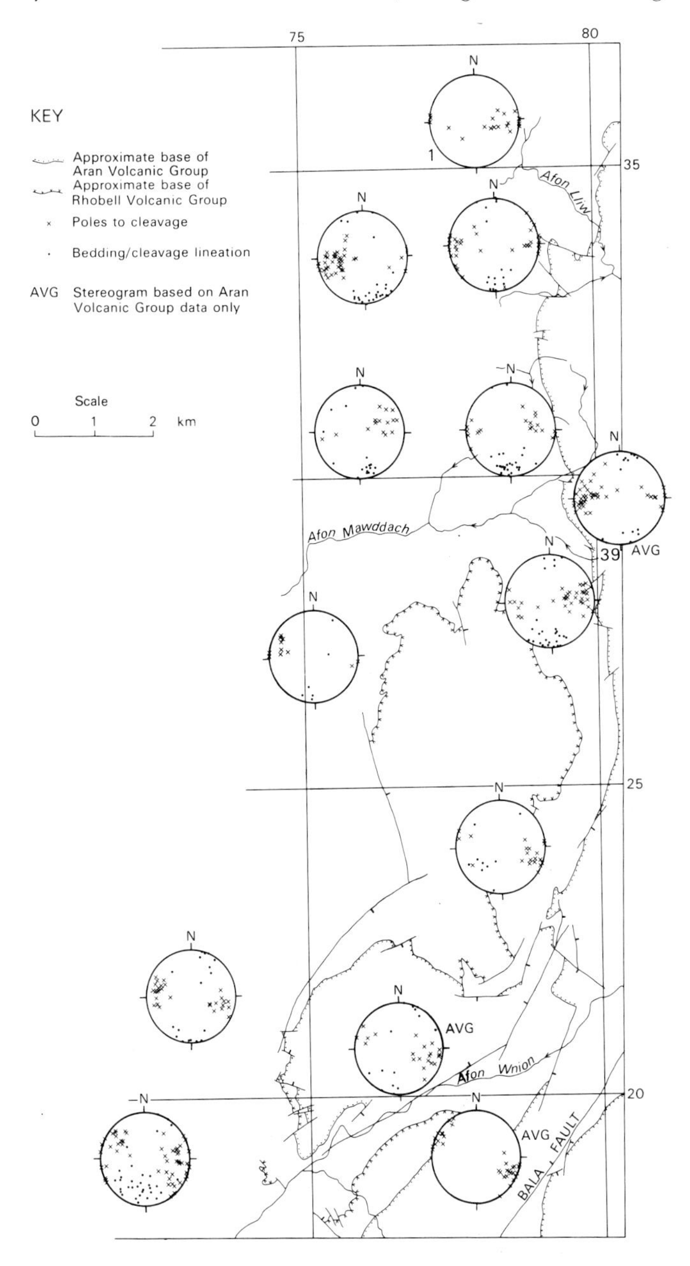

Figure 21 Stereograms showing poles to cleavage, and bedding/cleavage intersection lineations in the eastern Harlech Dome

the late Tremadoc folding, may have been accentuated during the later fold phase. Their N–S trend, which does not conform to the Caledonides trend, is clearly an inherited characteristic.

The several phases of folding seen within the eastern flank of the Harlech Dome are correlated with those of previous workers in Table 9. It is apparent that deformation under similar stress fields has taken place at different periods; correlation is, therefore, tenuous.

CLEAVAGE

A penetrative cleavage is only locally present. It is in many places clearly identifiable as an axial-plane cleavage, well developed in the hinge areas of minor folds and absent on their flanks. In many rocks the cleavage is defined by the reorientation of platy minerals. This is strongly evident in strata within the Harlech Grits Group. Elsewhere the cleavage is defined by fractures containing hematite or goethite, and within which sericite may be realigned. Cleavage is poorly defined or absent in coarse-grained sandstones, tuffs and tuffites. It is also generally absent in the coarse quartzose siltstones within the Mawddach Group, though in places preferentially orientated sericite is present in the matrix.

In the acid tuffs there is some evidence of realignment of phyllo-silicates cross-cutting eutaxitic textures. Within some pelitic beds, particularly in the Maentwrog, Cwmhesgen and the lower part of the Allt Lŵyd formations, there is evidence of a cleavage lying parallel to bedding. This is defined by the alignment of chlorite pellets and may be the result of loading on flocculated clay particles with a component of concentric shear (de Sitter 1964, p. 65). Where this bedding-plane cleavage, which may be the equivalent of the early 'bedding plane' cleavage described by Roberts (1967), is well developed, the main cleavage takes on the aspect of a crenulation cleavage.

The main cleavage trends 10 to 15° west of north in the north-eastern part of the district, and 10 to 15° east of north over most of the remainder of the re-mapped eastern strip (Figure 21). Locally a NW-trending cleavage (302°) is present. Dips range from 25°E through vertical to 25°W. Part of this range can be accounted for by the large angle of refraction of the cleavage, over 50° in places, where it passes from competent to incompetent beds, and the development of cleavage fans in many of the folds. Poles to cleavage planes plotted on stereograms (Figure 21) show a wide spread in both Ordovician and Cambrian strata. Cleavage/bedding lineations are predominantly southwards plunging within the Cambrian but perclinal in the Ordovician (Figure 21). To the north in the Migneint area, Lynas (1970) observed that cleavage/bedding lineations plunge towards the north-east. This variation may support the hypothesis that the cleavage was superimposed on an existing dome.

Although the evidence suggests that there is more than one set of cleavage, no direct evidence of the NW-trending cleavage having been folded during the later north-easterly phase has been recorded.

Ramsay (1866) recorded that the lower Palaeozoic rocks of North Wales are affected by two cleavages which he dated as 'Llandovery' and 'pre-Upper Old Red Sandstone'. Since then a number of cleavages have been described by authors, but the evidence for these is not always convincing. From detailed fieldwork Roberts (1967) described three end-Silurian cleavages in the Llwyd Mawr Syncline to the north-west of Harlech, and a similar pattern of deformation has been described from other areas in North Wales (Bassett, 1954; Helm, Roberts and Simpson, 1963; Fitches, 1972). While there is general agreement on these end-Silurian deformations several authors have argued that the prominent slaty cleavage developed early, in water-saturated sediments (Lynas, 1970; Davies and Cave, 1976). Lynas suggested that cleavage is rarely found below the Dolgellau Beds, and its development may be associated with the Caradocian volcanicity. Bates (1975) described sandstone dykes intruded along the cleavage in the Llanbedr Slates north of Barmouth, and concluded that this cleavage was also formed early in the tectonic sequence during dewatering and lithification and later enhanced by pressure solution and recrystallisation. Bromley (1971) dated the slaty cleavage in the aureole of the Tan y grisau granite near Blaenau Ffestiniog as post-Caradoc; he refuted the arguments for cleavage being a dewatering phenomenon, reiterating Shackleton's opinion that the slaty cleavage has been produced by one tectonic event. Bromley also contended that other deformations are usually much less intense, only recognised in relatively incompetent rocks, and usually confined to restricted areas or even particular stratigraphic units.

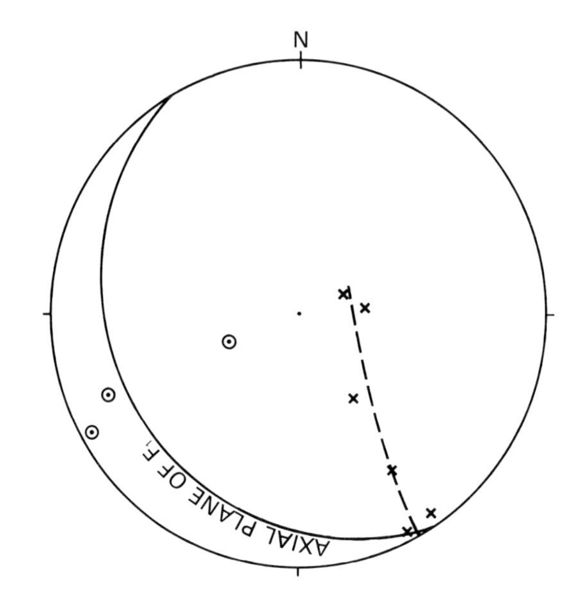

⊙ Pole to bedding on main fold

× Plunge of parasitic small-scale fold axes

Axial direction of F_2 approx 078°

Figure 22 Stereogram showing the reorientation of early folds by a later fold phase

In this district the evidence from the folding favours deformation as a continuous process, although the climactic end-Silurian event is responsible for the major regional structures. Within any stress system producing folds in a given rock sequence, a cleavage will develop when the strain conditions are appropriate and thus the cleavage may well be diachronous as suggested by Dewey (1969). AAJ

FAULTING

The fault pattern in this area (Figure 18) is complex, but it is possible to recognise a number of dominant trends and types of faulting and establish a relative chronology. In the following account convenient sub-headings have been employed as a descriptive framework, no formal classification is implied.

Reversed faults

On the western side of the district, the N-trending Moelfre Fault, with a westward downthrow, was described by Matley and Wilson (1946) as reversed. It is offset by a normal ENE-fault near Coed Crafnant [620 289]. Near Nant Budr [750 350] in the north-eastern part of the district a set of reversed faults, trending 350° and with a westward downthrow, is also offset by younger normal faults trending 075°. These reversed faults are parallel to minor folds in the adjacent rocks, the axial planes of which dip in the same sense as the fault surface. Bedding is steep adjacent to the faults and is locally overturned on the downthrow sides; where the rocks are strongly cleaved. Minor reversed faults with the same trend, but throwing down to the east, have been recorded in the Ordovician rocks in the upper part of the Afon Mawddach valley [802 285], and also trending 010° and throwing down to the east, in the upper Nant Ganol area [778 332]. Reversed faults with other trends are uncommon, but north of Buchesydd [780 211] the junction between the Ordovician and Cambrian rocks is marked by a fault with considerable throw. The trend of the fault is E–W and it appears to hade to the north, thus giving the fault a reversed throw. Eastwards the fault passes under boulder clay, and westwards it seems to end against a NE-trending fault.

PMA, AAJ

NNE faults

In the north-west a belt of normal faults with individual trends between 020° and 040°, and throwing down either to the east or west, obliquely traverses the Traeth Bach Syncline. Some of the faults were named and described by Matley and Wilson (1946) who designated them, together with the NNW- and N-faults, as meridional. This classification, however, is an oversimplification. The belt dies north-north-eastwards and does not appear to cross the Vale of Ffestiniog. Subsidiary faults within the belt trend in a variety of directions, though there is a scarcity of NNW faults.

Several NNE- to NE-faults, generally parallel to the bedding strike, occur in the south-east. They include the major Bala, Dolgellau (also called Wnion or Ceunant), Derwas/Afon Wen, Llyn Cynwch, and Llanfachreth faults, together with many minor structures. Apart from the Bala and Dolgellau faults, which are considered separately, all the major faults appear to be normal with a downthrow to the south-east. The westernmost of these, the Derwas/Afon Wen Fault, can be traced through Ordovician rocks on the north-eastern slopes of Cader Idris (Cox and Wells, 1921) into the Mawddach valley south of Llanelltyd. From there north-north-eastwards along the Mawddach and Wen valleys the existance of the fault is speculative, but this line marks an important structural divide which may mark the hinge of the pre-Rhobell Volcanic Group tilting and folding, with moderately dipping beds on the west and steeply dipping or overturned beds on the east. In the blocks between the major faults there are many dip faults with small throws, mostly trending between 300° and 320°, and with downthrow direction more or less evenly divided between south-west and north-east.

NNW faults

The major faults on the north-eastern side of the Harlech Dome trend between 320° and 345°. They appear to be normal faults throwing down either westwards or eastwards. Matley and Wilson (1946) named and described three of them, the biggest of which is the Trawsfynydd Fault. Near Trawsfynydd this fault has a downthrow to the east of about 1200 m. It is exposed in Gwynfynydd gold mine where the crush zone is about 1 m wide; slickensides on the wall confirm the normal downthrow to the east. In the mine and the adjacent area there are several parallel faults and also a suite of westward-dipping minor faults with the same trend. The main fault displaces the gold lodes and earlier ENE faults in the mine. Northwards the Trawsfynydd Fault is probably continuous with the Cwm Bowydd Fault which crosses the Tan-y-Grisiau granite. To the south it ends in the Capel Hermon area near the Derwas/Afon Wen Fault.

The Bwlch Goriwared Fault on Rhobell Fawr also appears to end against the NNE fault-zone.

While in most areas there is evidence of later movement, NNW faults appear to predate the intrusion in the late Tremadoc of the Craiglaseithin [735 325] microtonalite laccolith; on the south-eastern side of Rhobell Fawr, faults with this trend have a greater throw in the Cambrian than in the Rhobell Volcanic Group, suggesting that they were initiated prior to the eruptions. In this area (i.e. south-east of Rhobell Fawr) these faults are displaced by NE-trending faults, and a sill-like intrusion of dolerite near Blaenau crosses one of them [793 229]. Several of these NNW-trending faults on Ffridd yr Allt Lŵyd may have been active during the deposition of the Ordovician sediments and volcaniclastics.

North faults

Faults with this trend are uncommon. The main one exposed is the Caerdeon-Bodlyn Fault which nearly coincides with the axial plane of the Caerdeon Syncline. Smaller parallel faults occur in the culmination of the southward plunging end of the Dolwen Pericline to the east of Bontddu, and one of them displaces the Clogau/St David's gold lode. Another trends north from Rhobell Fawr and is displaced by WNW- and ENE-trending faults. It is essentially a strike fault, progressively cutting out more of the Cwmhesgen Formation as it passes northwards. The northern portion, north of Afon Lliw, is concealed under boulder clay.

Tension faults

In an area defined by the Moelfre and Upper Artro faults to the west and the Trawsfynydd and Derwas/Afon Wen line to the east, minor normal faults follow two main complementary directions. On a rose diagrams (Figure 23) the maxima are 330° and 060°. There are approximately equal proportions of the two sets in the two quadrants; for each set the direction of throw is evenly divided between the two, opposing possibilities. Faults with these trends are not confined to this area but within it they are dominant. They are most common on the flanks of the Dolwen Pericline and scarce along the crest.

The two main directions of faulting coincide with the trends of dykes and mineralised quartz veins in the area (Figure 23). The majority of dykes trend between 320° and 330°, and quartz veins show a maximum at 060° with a variation between 025° and 075° in common with the faults. Both dykes and quartz veins commonly follow faults, a good example of the former being near Clogau mine [6753 2050]. It is likely, therefore, that the faults are tension fractures.

During folding tension fractures form perpendicular to the principal stress in antiforms and along it in synforms. In this area the main fault directions make equal angles to the axis of the Dolwen Pericline and are not readily explained by any stress system involving horizontal compression. They are more likely to be the result of vertical movement, possibly initiated during the formation of the Dolwen Pericline and reactivated during the period of relaxation at the end of the end-Silurian deformation

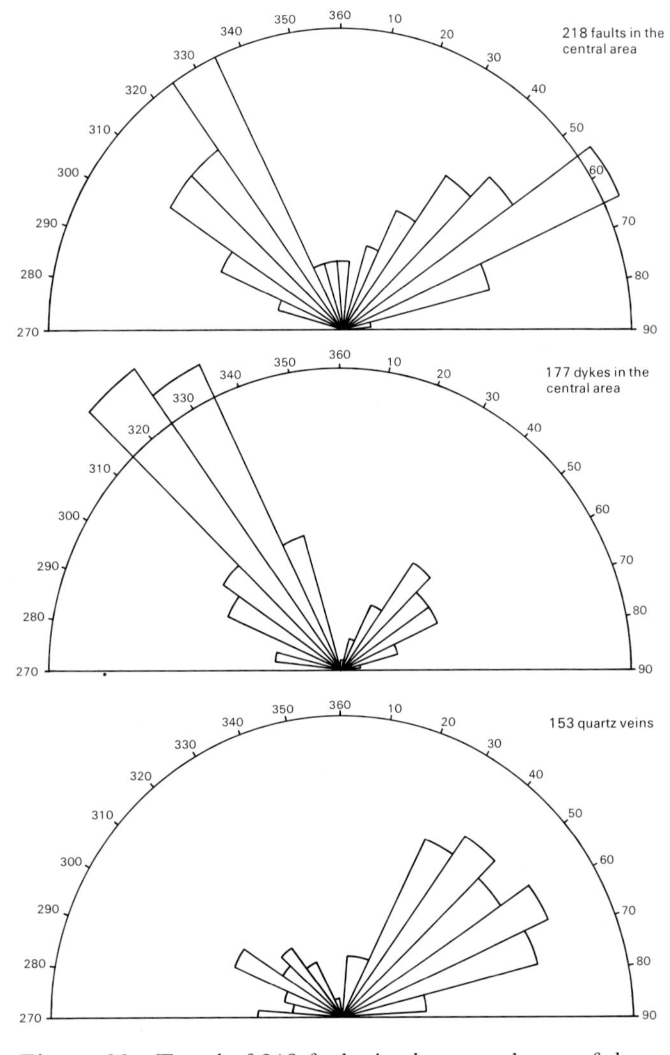

Figure 23 Trend of 218 faults in the central part of the Harlech Dome compared with the trends of major quartz veins and dykes

Faults related to folding

There are many small faults in the area which do not conform to the trends of the major faults. In some cases they can be directly related to local fold structures. A set of NNW-faults in the hinge zone of the Llafar Anticline are possibly crest faults. The Foel Boeth laccolith [785 345] which domed the overlying sedimentary rocks is cut by a set of NW cross-faults, which are subparallel to a complex set of cross-faults in the folds around Blaen Lliw to the east. PMA

Faults active during Ordovician sedimentation

In general these faults trend NE and WNW, and are made apparent by a dramatic change in thickness of some strata across the line of the fault. The effect of such penecontemporaneous faulting can be seen in most of the formations, but is perhaps best illustrated in the Allt Lŵyd Formation where the lower, more argillaceous facies, which dies out to the south, changes in thickness across faults [8030 2563 and 8018 2754]. The thick slumped unit [800 292] on Ffridd yr Allt Lŵyd is confined on the west by a NNW-trending fault and on the south by an ENE fault. It is overlain by an anomalous sequence consisting of a very thin crystal tuff overlain by hyaloclastite. The fault pattern suggests that the mudflow was deposited in a fault-bounded basin or graben which was gradually filled by tuffs of the Benglog Volcanic Formation. AAJ

Bala Fault

The Bala Fault is one of the major structural features in North Wales, and can be traced from Tywyn (Jehu, 1926) through the south-eastern corner of the district to Bala. According to Alder (1976) it does not link up between Bala and Corwen with the Bryneglwys Fault. Wells (1925), Jehu (1926) and Shackleton (1952) all described it as a sinistral wrench fault. Bassett, Whittington and Williams (1966) described a precursor to the Bala Fault in the area around Bala, which they claim was active in pre-Ashgillian times and was oblique-slip with downthrow to the south-east and dextral slip. Both Rast (1969) and Ridgway (1976) produced evidence for pre-Arenig movement of a similar style in the area south-east of Bala.

The fault surface is nowhere visible in the district.

However, despite extensive drift cover, exposures in Nant Helygog [7920 1890], near Pont Rhyd-y-gwain [7984 2012], and in the area between, indicate a dip separation across the fault on a presumed vertical plane in the order of 3.5 km, with a downthrow to the north-west. The fault trends 035°, and is subparallel to steeply dipping beds. In the adjacent parts of the Aran Mountains, Dunkley (1978) found evidence to dispute the claims of Rast (*op. cit.*) and Ridgway (*op. cit.*). He concluded that the throw across the Bala Fault was down to the north-west, and that the most recent movement along it was sinistral and about 5 km. PMA

To the south-east of Rhobell Fawr a number of faults occur which are subparallel to the Bala Fault. A second set of faults trend at about 305° and 340°; these show near horizontal slickenside indicative of oblique slip movement.

The Dolgellau (Wnion or Ceunant) Fault, which follows the Wnion valley (ENE) across the district, may be an offshoot of the Bala Fault (Wells, 1925) though it appears to be broken by NE-trending faults which are parallel to the Bala Fault. AAJ

Mochras Fault

Seismic studies (Griffiths, King and Wilson, 1961; Blundell, King, and Wilson, 1964; Bullerwell and McQuillin, 1969) have clearly demonstrated the existence of a flat-lying low-velocity succession in Tremadoc Bay, thus corroborating the earlier gravity results of Powell (1956). A steep interface (i.e. greater than 45°) between these rocks and the high velocity Lower Palaeozoic rocks of the hinterland has been suggested. This interface has been termed the Mochras Fault, although there is no direct geophysical evidence to indicate faulting. The Mochras Farm Borehole, 3 km to the west of the known Cambrian outcrop, penetrated 1.9 km of Mesozoic and Tertiary sediments. By extrapolation of the IGS seismic reflection profile 2B to the Mochras Farm Borehole and the adjacent fault, a tentative depth of 2.7 km to the base of these sediments is suggested for that area (Figure 24).

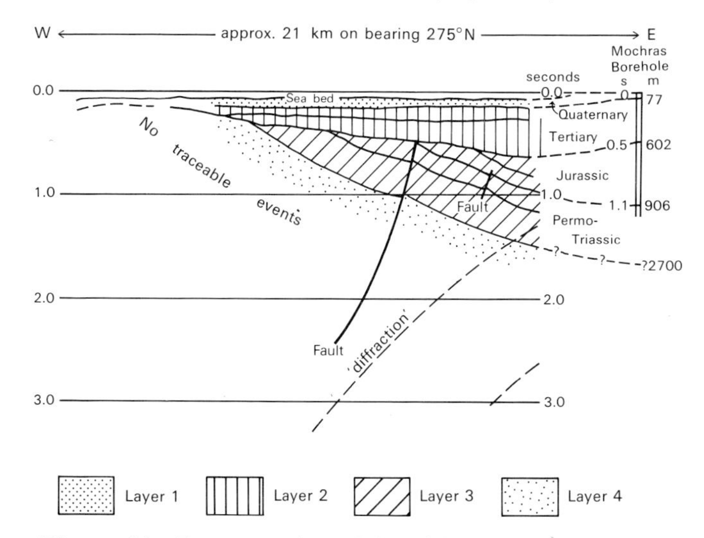

Figure 24 Interpretation of the IGS seismic reflection profile (2B in Bullerwell and McQuillin, 1969). The relationship to the layers deduced from refraction methods and to the Mochras Farm Borehole are indicated. The vertical scale is in seconds

The linear Bouguer anomaly gradient which dominates the gravity field near the coast has been interpreted by many authors as the contact between the low density sediments beneath Tremadoc Bay and the Cambrian rocks of the Harlech Dome. In particular Blundell, Davey and Graves (1971) presented a cross-section (their fig. 12), incorporating the seismic reflection data of Bullerwell and McQuillin (1969), which shows step-faulting along the eastern margin, with a total throw of more than 2000 m.

Z. K. Dabek (personal communication) has measured nine detailed gravity traverses in the area of the steep gradient, in order to determine the trace of the fault. These results have been incorporated in the published 1:50 000 Geological Sheet 135. Dabek observed that the anomaly decreases to the north, and the trace of the fault becomes more difficult to identify in this direction. This is caused by thinning of the Mesozoic and Tertiary sediments in the basin, possibly related to a decrease in the throw of the fault. In the south of the area which he studied (4 km N of Barmouth), the basin is deduced to be no more than 4 km deep. To the north of Harlech, the fault anomaly interferes with the anomaly which Blundell, Griffiths and King (1969) deduced is caused by a buried river channel (Figure 25).

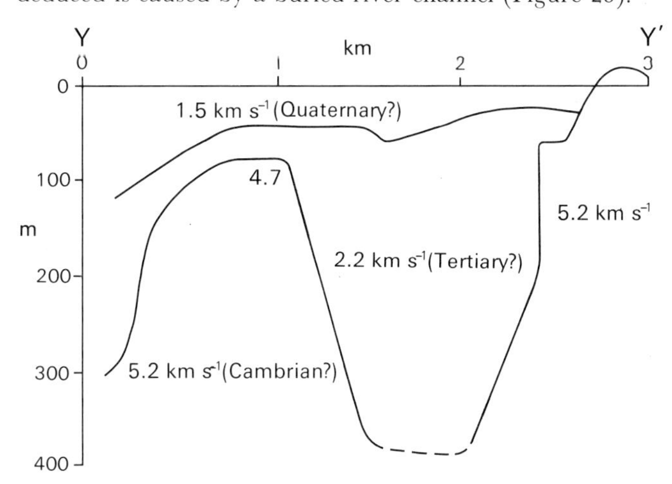

Figure 25 An interpretation of the seismic results (compatible with gravity data) along Y-Y′ (Figure 27) (after Blundell and others, 1969)

The Bouguer gravity anomaly has been calculated for a two-dimensional model across the fault and incorporates data from the borehole and seismic profiles. The results are shown in Figure 26, where they are compared with the residual anomaly obtained by removing a linear regional field (based on values over Cambrian outcrops on the Lleyn and Harlech Dome away from the influence of the fault anomaly) from one of Dabek's traverses. There is freedom in the model to adjust the fit by selection of the Permo-Triassic density, by the dip of the base of the Permo-Triassic beneath the borehole, and by varying the hade of the fault. The density selected satisfied the amplitude of the anomaly a short distance offshore, but may vary laterally as the margin of the basin is approached. A 50° hade for the fault-plane and a westward dip on the base of the Permo-Triassic optimise the fit. On this basis the vertical throw of the fault is at least 2000 m, and there is no evidence for or against step faulting.

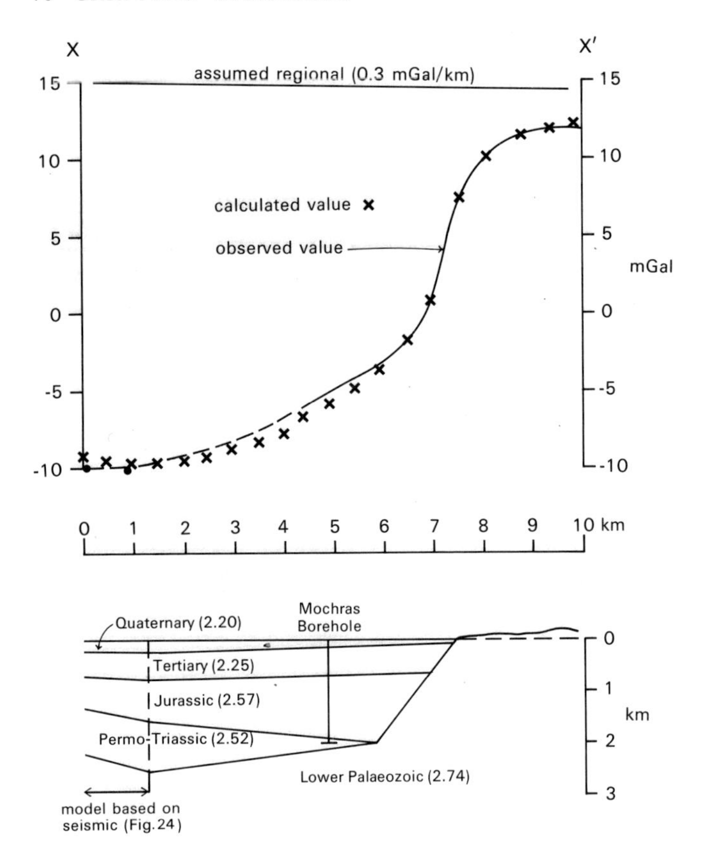

Figure 26 Bouguer gravity anomaly along profile X-X′ (Figure 27) over the Mochras Fault, compared with values calculated for a model based on seismic and borehole data. (Density values in mgm^{-3})

Despite many adjustments of the model for the distribution of magnetic rocks, it did not prove possible to reproduce closely the aeromagnetic anomaly adjacent to the fault. Reasons for this are the limited validity of the assumption of two-dimensionality, 'aliasing' of the anomaly shape, the possibility of remanent magnetisation, and the complex (and uncertain) distribution of magnetic materials. The combination of thickness, susceptibility and hade shown on Figure 30 (p. 75) optimises the model, although others may be valid. A shallow fault-hade (approximately 45°) has been shown, but a precise estimate is not possible. An improved fit to the magnetic profile to the west of section X-X′-X″ is achieved if the top of the magnetic rocks dips westwards from the fault for at least 7 km. Beyond that it may dip in the opposite direction, so as to crop out on the Lleyn; the anomaly is not sensitive to such variations. To the west of the fault, the magnetic material is at 4 km depth. By analogy with rocks at outcrop, these magnetic rocks at depth may be Lower Cambrian. The non-magnetic interval of 2 km between the base of the Permo-Triassic and the top of the supposed Lower Cambrian rocks must consist of rocks with a density of 2.74 Mg m^{-3} to satisfy the gravity anomaly. These are probably Upper Cambrian and Ordovician mudstones.

If the assumption and deductions made above are valid and the 2.5 km thickness of the Lower Cambrian rocks found in the Harlech Dome continues beneath Tremadoc Bay, then a simple trigonometrical solution gives a vertical throw for the fault at Mochras of approximately 6 km. The planar interface resulting from the geophysical interpretation is probably a simplification of the faulting; it may represent the resultant of a number of faults, and subsequent erosion and subsidence.

The presumed fault continues to the north under Morfa Harlech, which was studied by Blundell and others (1969) using a combination of seismic refraction and gravity methods. They were able to deduce a buried valley about 380 m deep, cut into ?Cambrian rocks with a velocity of about 5 km s^{-1}, and filled with Tertiary rocks with a velocity of 2.2 km s^{-1} (Figure 25). The possible continuation of this valley out into Cardigan Bay was noted on sparker records where it may be represented by a synclinal structure. However its gravity expression does not extend much beyond the coastline (fig. 7 *in* Dobson, Evans and Whittington, 1973), although the coverage is not adequate in the critical area adjacent to the coast. The valley may mark a possible continuation of the Mochras Fault, although its precise position is a matter of speculation. If the major movements of the fault plane were pre-Tertiary, then the fault may run up the axis of the valley, which might have formed along the resultant zone of weakness; if they are Tertiary in age then the eastern side of the valley may mark the fault-plane. The symmetrical cross-section of the valley suggests pre-Tertiary movements, although (because of the different elevations of Cambrian rocks on either side of the fault) later movements cannot be ruled out.

The negative Bouguer anomaly associated with the buried valley extends northwards into the Glaslyn valley and marks the course of the ancient Afon Glaslyn. The buried valley at Morfa Harlech is evidence of at least 380 m of uplift of the hinterland in relation to sea level, causing rejuvenation of the river with consequent down-cutting, followed by subsidence to the present position. IFS

The existence of this valley is further supported by the interpretation that the Tertiary sediments in the Mochras Farm Borehole are subaerial fluvial deposits, and O'Sullivan (1979) does not rule out the possibility that the Mochras Fault was active during their deposition. He suggests that the conglomeratic beds from the base of the Tertiary sequence may be the product of flash floods, derived either from an active fault-scarp or more likely (as the constituent cobbles are well rounded) from an older uplifted surface. The finer grained sediments which dominate the upper part of the sequence are interpreted as point-bar and flood plain deposits. Thus the fault-controlled scarp marked the edge of the Tertiary basin of deposition with a channel draining southward.

Herbert-Smith (1979) assigns a Middle Oligocene to early Miocene age to the Tertiary deposits whilst Wilkinson and Boulter (1980) suggest a wholly Oligocene age. The thick sequence of Jurassic rocks is entirely fine-grained and shows no evidence of underlying a marginal facies; this leads to the conclusions that the Lower Palaeozoic rocks were deeply buried at this period, and that the main movement on the fault is post Jurassic and pre-Oligocene although movement may have continued into the Oligocene. AAJ

CHAPTER 6

Geophysical investigations

Geophysical techniques have been instrumental in advancing the understanding of the geology of the Harlech district. A variety of methods has been applied on land and the adjacent offshore area to obtain both regional cover and data over specific targets. In this chapter the resulting information, which demonstrates the relationship between the geophysical data and the geological structures, is reviewed.

The following sections cover gravity, magnetic, seismic and electromagnetic surveys, in each case presenting both the results of published work and some new interpretations. In addition, Appendix 2 describes and lists selected physical properties of rocks in the district.

GRAVITY SURVEY

Regional gravity coverage of the Harlech district was obtained by Powell (1956) who measured about 50 stations in the area and delineated the principal gravity features. Further coverage by the Geological Survey in 1974–75 and 1980 resulted in approximately one station per 2 km^2 of the land area. In the offshore part some adventurous 'land' stations were established by Griffiths, King and Wilson (1961) on Sarn Badrig (the sand bar reaching out into Tremadoc Bay). Blundell, Davey and Graves (1968) and Dobson, Evans and Whittington (1973) both reported the results of surveys using marine gravity meters. Detailed gravity traverses have been measured by Griffiths and others (1961) and the Geological Survey over the Mochras Fault, and by Blundell, Griffiths and King (1969) over buried channels in the Mawddach and Glaslyn estuaries.

These data are compiled and presented here as a Bouguer gravity anomaly map (Figure 27). Data incorporated on this map are adjusted to National Gravity Reference Net, 1973 and reduced using International Gravity Formula, 1967, and a standard density of 2.67×10^3 Mg m^{-3}. Previously published data relate to the outdated Potsdam system; to compare them with modern data, approximately 2.9 mGal should be subtracted from the old values.

The most eye-catching feature of the Bouguer gravity anomaly is the steep gradient in the coastal plain and surrounding hills between Harlech and Barmouth. The gradient averages 2.5 mGal km^{-1} (Bouguer anomaly decreasing westwards) over a belt 6 km in width, but the steepest part is about 12 mGal km^{-1}. The coastal anomaly extends to the north into Morfa Harlech, where it interferes with a negative anomaly described by Blundell and others (1969). The interpretation of these anomalies is described in the section on the Mochras Fault in Chapter 5.

Bouguer anomaly values increase gradually to the east, with several broad, low amplitude anomalies superimposed. These anomalies show only weak correlation with exposed structures. There is a relative minimum over the Rhinog Formation in the Caerdeon Syncline [65 20 to 64 30]. The relatively low density of the grits that form the high ground of the area may have caused errors in reduction of the data.

A weak W-trending minimum overlies the Dolwen Pericline [72 31], which conflicts with the data presented by Powell (1956); the conflict may result from poor coverage on both maps. A N-S anomaly [75 28 to 75 20] correlates with the denser argillaceous Mawddach Group and associated igneous rocks.

A NW-dipping gradient, which is parallel to the Mawddach along the southern margin of the district, is a result of dense rocks beneath Cader Idris. Two salients, near Llanelltyd [71 19] and Brithdir [78 19] may be related. The latter is probably caused by dolerite intrusions; the former has no exposed explanation.

On the northern margin of the district Bouguer anomaly values increase across the E–W gradient of c. 1 mGal km^{-1} into the area of the Snowdon Syncline. Powell (1956) has related this to crustal thinning.

MAGNETIC SURVEY

Magnetic total field coverage of the Harlech district was obtained during the aeromagnetic survey of the United Kingdom (Geological Survey, 1965). The area was surveyed along N–S flight lines spaced 2 km apart, with a mean terrain clearance of 1000 ft (about 300 m) and with tie-lines at 10 km intervals. Magnetic anomalies were related to a standard planar reference field. The resulting field for the area of the Harlech district and its immediate surroundings is presented as Figure 28.

Detailed low-level aeromagnetic coverage of the eastern part of the district was obtained during the Mineral Reconnaissance Programme carried out by the Geological Survey for the Department of Industry (Allen, Cooper and Smith, 1979). The survey was flown along E–W flight lines (except in the southern third where NW–SE oriented lines were measured), with a separation of 200 m and mean terrain clearance of 150 ft (about 45 m). The same reference field was used for this survey as for the national survey. The surveyed area is indicated on Figure 28. Figure 29 shows 10 km grid square SH 72 from this survey, simplified by omission of some contours.

Information gathered from the Bryn-teg Borehole (Smith and McCann, 1978) and during the ground follow-up of the low-level aeromagnetic survey (Allen and others, 1979) has enabled the sources of many of the anomalies to be identified. Sites at which these ground surveys have been carried out are shown on Figure 28.

Broad similarities may be seen when the airborne and ground surveys are compared, although several factors outlined below cause differences in detail:

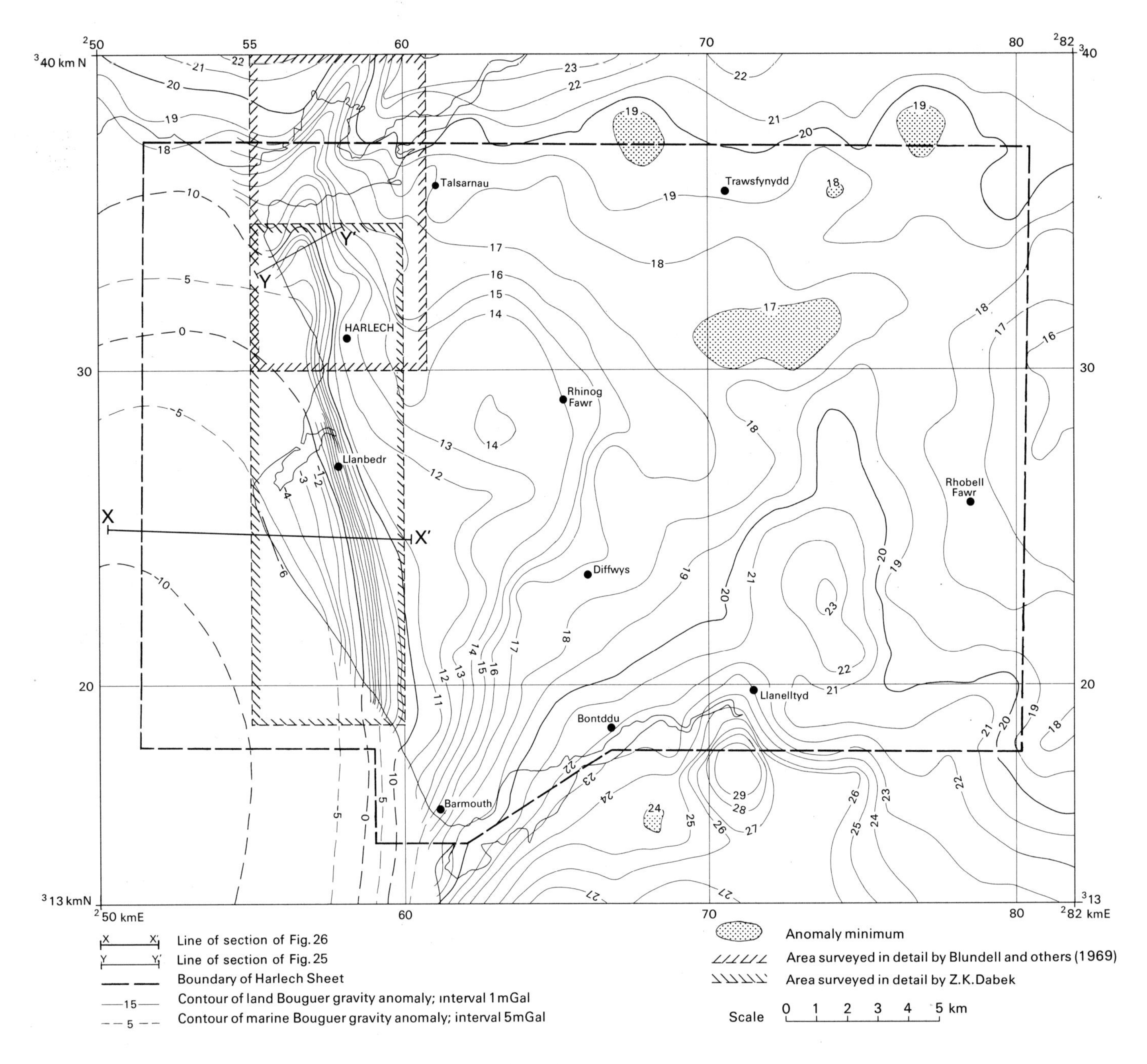

Figure 27 Bouguer gravity anomaly map of the Harlech district and surrounding area

1 Over much of the area magnetic rocks are close to the surface. As the distance above the causative body is increased two effects are noted. Firstly, the high frequency components of anomalies reduce relative to the lower frequencies, so that higher-level surveys produce broader anomalies. Secondly, the amplitude of an anomaly decreases according to the inverse power laws so that, though the amplitude of an anomaly measured during the two airborne surveys may only differ by a small amount, the amplitude on the ground may be up to an order greater.

2 In these latitudes the position of a central maximum of an induced magnetic anomaly will remain substantially the same at whatever level it is observed; the flanking minima, however, may shift considerably and interfere with adjacent anomalies in different ways.

3 'Aliasing' of anomalies commonly occurs because of sampling at intervals which may not represent the wave-length of the anomaly. This may be particularly important on linear anomalies which parallel the flight direction, as for example anomaly C-C′ on Figure 28.

4 The process of producing small-scale maps of magnetic intensity involves several stages which result in loss of detail, so that small anomalies and trends, which may be recognisable on the flight records or original full scale maps, are not seen on the final product.

In the following text, anomalies identified by letters on Figure 28 are described and related to the low level survey and any relevant ground measurements made over them. The principles of this description are shown by the cross-

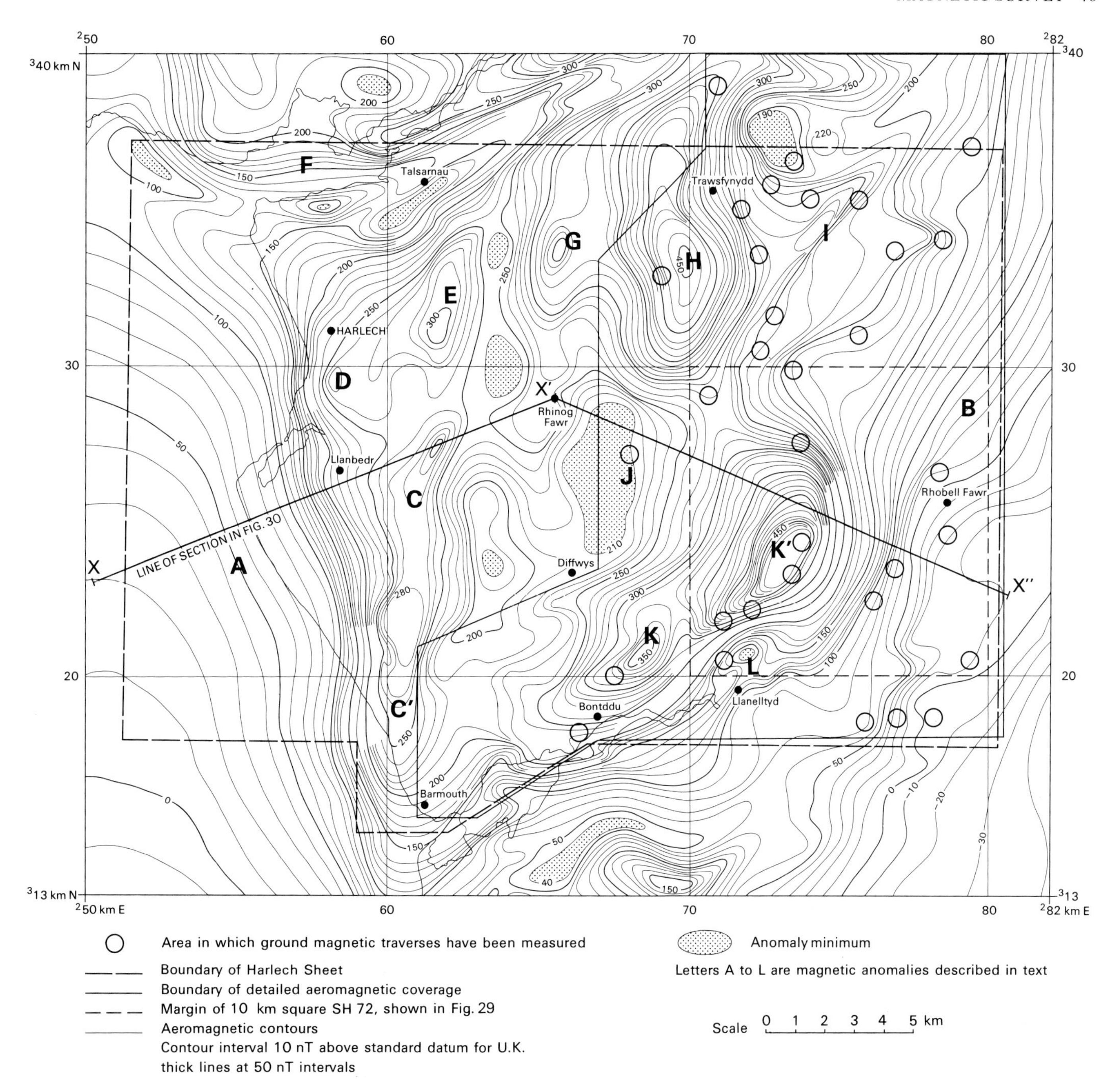

Figure 28 Regional total field aeromagnetic map of the Harlech district and surrounding area

section X-X′-X″ (Figure 30), which follows the same general line as the cross-section on the geological map.

The magnetic character as seen on Figure 28 falls into three distinct divisions: a central rhombic area, in which the magnetic gradients are steep and intensities range from about 0 nT to about 500 nT, and the surrounding areas, (A and B on Figure 28) which continue east to the Cheshire Basin and south and west to St David's Head, in which gradients are low and the field strength is close to the regional value. The form of the field in areas A and B is due partly to relaxation of the field away from the anomalous area, which causes flanking minima, and partly to a substantial cover of weakly magnetic rocks so that the magnetic material is at great depth, resulting in anomalies with low frequency and low amplitude only. In Area A, over Tremadoc Bay, a possible interpretation shows the magnetic (presumed Cambrian) rock faulted down to 4.5 km and dipping westward. In Area B the Lower Cambrian dips east under the Ordovician rocks to a depth of 6 km at a point 6 km E of the end of the cross-section. Area B is also clear on the low level survey (Figure 29). Broad features on Figure 28 [78 24 to 78 19] are resolved in the low-level survey into many anomalies with amplitudes of up to 200 nT. For example, over the Rhobell Volcanic Group [78 24], the anomalies are discrete and show flanking

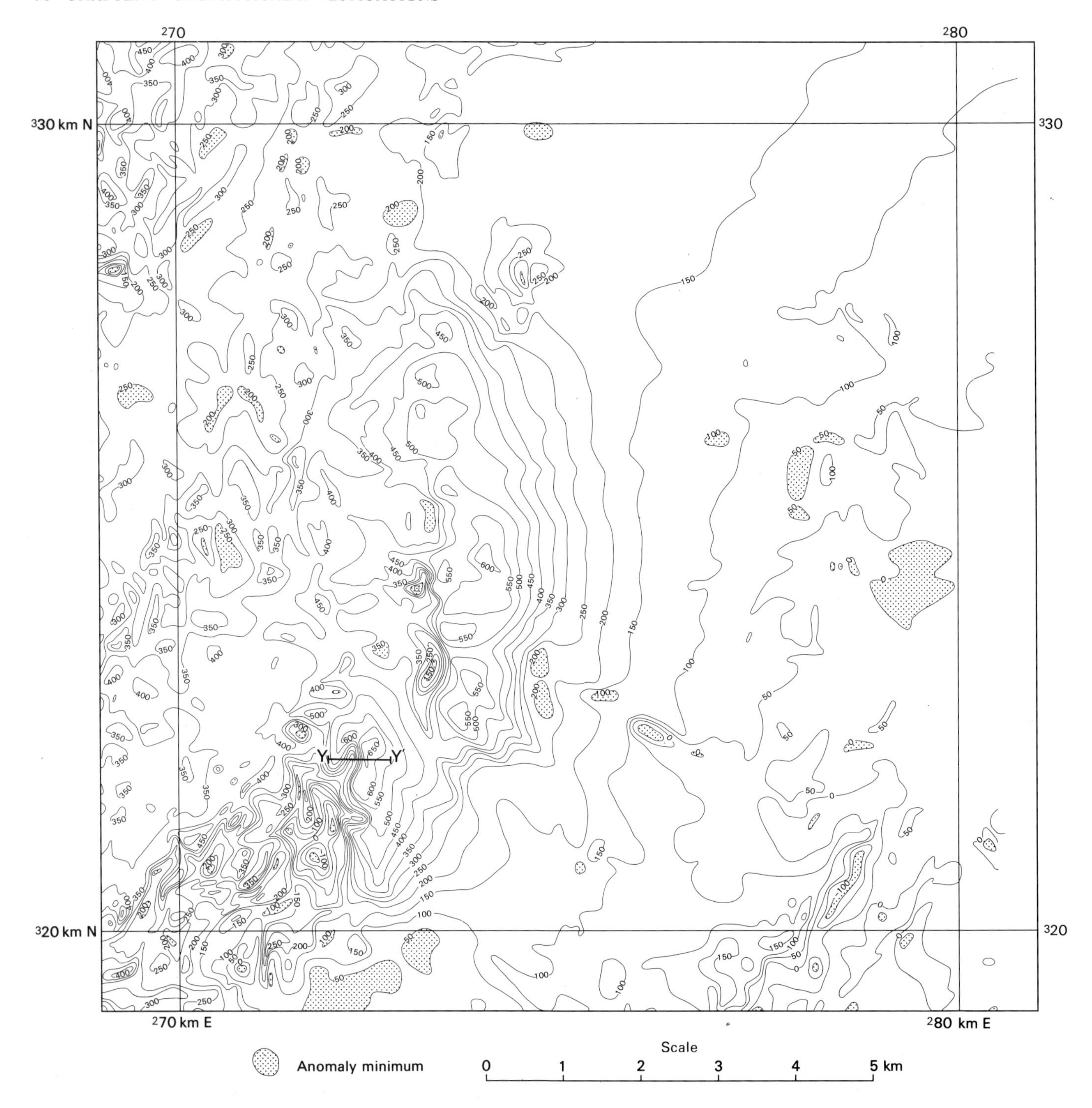

Figure 29 Detailed total field aeromagnetic map of 10 km grid square SH 72. Contour interval 50 nT, above standard datum for U.K. Y-Y′ is the profile illustrated on Figure 31

minima. Some correlate with faults, and may be due to steps at the base of the magnetic rocks. The interpreted susceptibility is 5-10 × 10^{-3} SI units and the rocks are normally magnetised. Elsewhere, the anomalies over dolerite intrusions [78 19] tend to interfere, but the edges and trends of some of the larger bodies are clearly indicated [77 19 to 79 21]. The amplitude and shape of the anomalies suggests reversed magnetisation with an intensity of magnetisation about 1500 nT. A linear NW-trending anomaly [76 22] also shows reversed polarisation and may be a concealed dolerite dyke-like body (Figure 29).

The central rhombic area is bounded by approximately linear belts of steep magnetic gradient. The steepest gradient on the western boundary marks the position of the plane of the Mochras Fault. Simple depth rules (Peters, 1949) indicate a maximum depth of 0.75 to 1.5 km to the top of the step, although these estimates are likely to be too high, because of 'aliasing'. The eastern margin of the area marks the position of the non-magnetic Middle and Upper Cambrian and Ordovician rocks of area B. In the south-east corner, and extending beyond the district, the Bala Fault marks the edge of the anomalous region, with steep-sided anomalies over the volcanic rocks of Cader Idris [69 13]. There is, however, a marked gradient, which decreases southwards across the Mawddach estuary. It separates the Cambrian succession (in the north) from the Ordovician (in the south). The north-western boundary of the anomalous area is not clearly defined because of the proximity to the north of the Snowdon Volcanic Group and its associated magnetic anomalies. However, steep gradients follow the Dwyryd estuary, where Harlech Grits are faulted against Mawddach Group and Ordovician shales.

The linear Anomaly C-C′ is the culmination of the gradient associated with the Mochras Fault, and lies over the outcrop of the Llanbedr and Rhinog formations and the concealed Dolwen Formation. The Rhinog and Dolwen formations both contain magnetite disposed in bands of high susceptibility (Smith and McCann, 1978; Allen and others, 1979). The truncation of these formations by the Mochras Fault is considered to cause the anomaly. To model the anomaly precisely, however, requires more geological control than is available. A fair comparison between the observed field and theoretical anomaly is shown in Figure 30, but improvement could result from including a highly magnetic core at a depth of (say) 4 km. Anomaly D probably has a similar geological cause. Although Anomaly E lies on a similar trend, it is not related to anomalies C-C′ and D, but falls over faulted Gamlan Formation and a suite of intrusive rocks. This western group of anomalies is truncated to the north by a complex flanking minimum, related to a suite of faults running up the Dwyryd valley. The western margin of Anomaly F, which lies over the dolerites of the Portmadoc area, is a prominent S-striking gradient and possibly represents a continuation of the Mochras Fault.

East of Anomaly C-C′ is a minimum which may be interpreted as being partly due to the less magnetic rocks in the Caerdeon and Foel Ddu synclines, and partly a flanking minimum to the Mochras Fault anomaly. Anomaly G occurs over the magnetic Rhinog Formation, within which some minor intrusive rocks are observed. Where these intrusions are covered by the low-level survey [68 34], weak anomalies subparallel to their strike may be seen. It is, therefore, possible that these intrusions are magnetic and contribute to Anomaly G.

One of the strongest features in the district is Anomaly H, with an amplitude of 150 nT. It is related closely to the outcrop of the Dolwen Formation in the centre of the Harlech Dome, although Smith and McCann (1978) provided evidence which showed that the known thickness and susceptibility are insufficient to explain in full the amplitude of the anomaly. They propose a core of magnetic material to the Dolwen Pericline. To the east of Anomaly H is the NE-

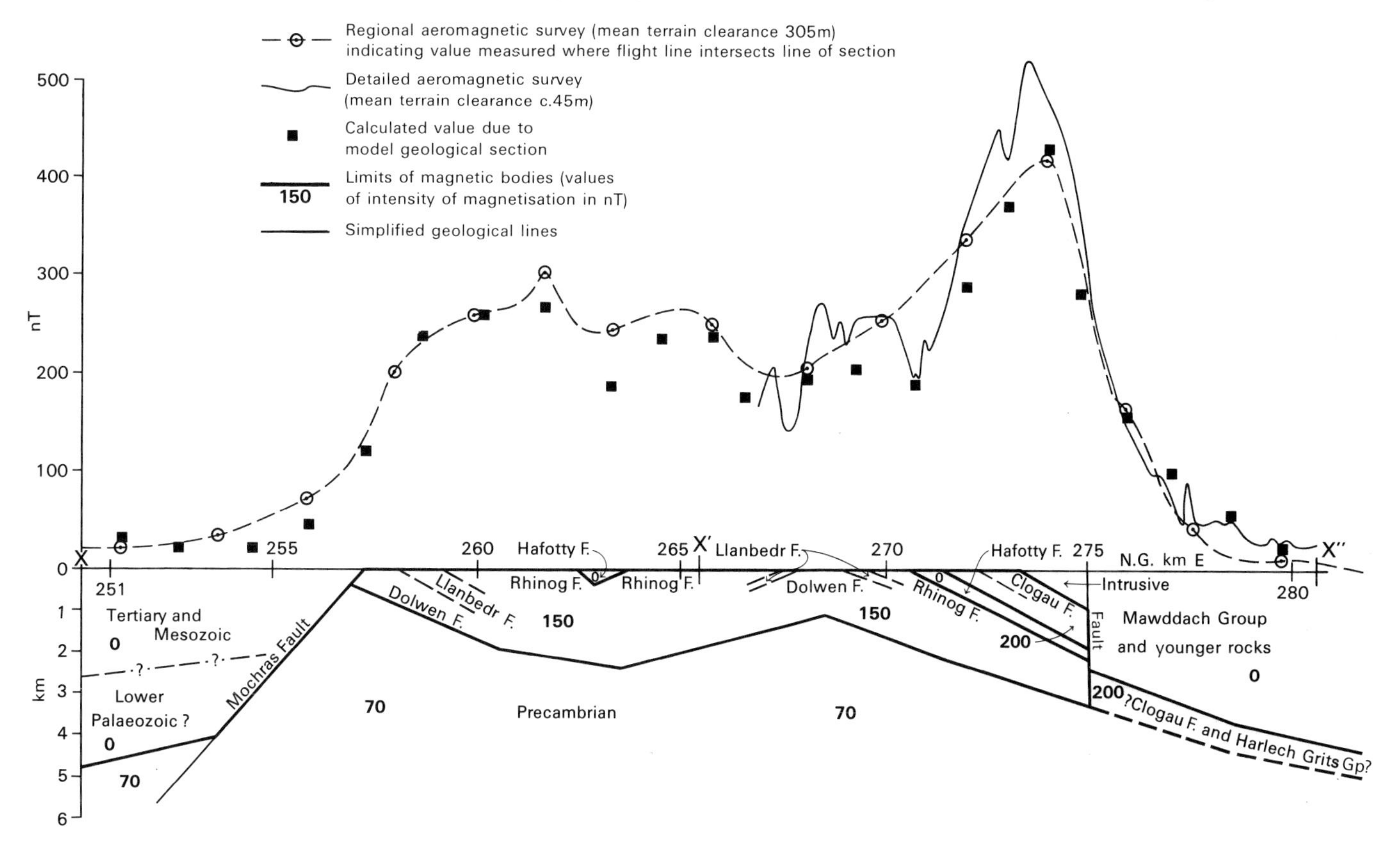

Figure 30 Profile X-X′-X″ (Figure 28) showing regional and detailed aeromagnetic data and a possible geological interpretation, incorporating know· geological data

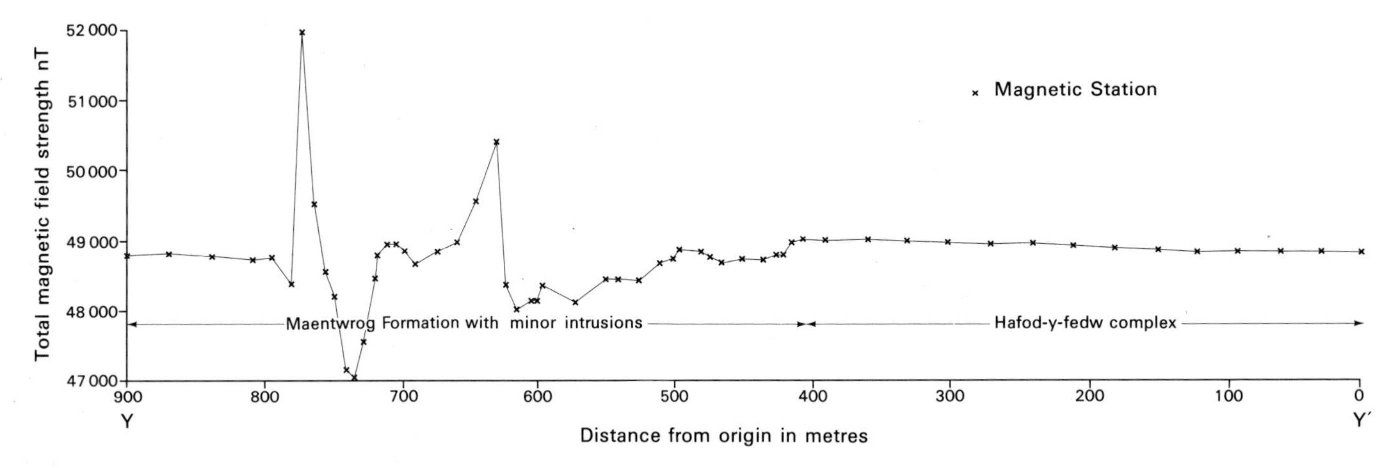

Figure 31 Total magnetic field strength measured on the ground along profile Y-Y′ (Figure 29) (after Allen and others, 1979)

trending Anomaly I, which corresponds to localised high amplitude and frequency features on the low-level survey. No clear geological origin of the feature has been determined. Although ground studies showed strong anomalies which correlate with the outcrop of the Clogau Formation, this does not adequately explain the greater extent of the whole feature, which is probably caused by over-ambitious contouring of the data. To the south-west of Anomaly H is a minimum J, which is probably a result of relaxation of the field away from maxima to the east and west. However, the low-level survey resolves the zone into separate small anomalies, due to local concentrations of magnetite in the Rhinog Formation.

Anomaly K-K′ is a SW-trending elongated maximum which is prominent on both aeromagnetic surveys. It has been investigated at several sites on the ground. The smooth eastern flank is shown on both surveys, and further demonstrated by the ground survey (e.g. profile Y-Y′ Figure 31, the location of which is shown on Figure 29). The steep western flank and the extension to the south-west are very different, however, and are shown with increased detail to be composed of a complex of high frequency anomalies, with changes of up to 3000 nT in 30 m and typical wavelengths about 50 m (Allen and others, 1979). Anomaly L appears as a simple minimum on the regional survey, but is shown to be more complex by the increased resolution and definition of the low-level survey.

An initial inspection of Anomaly K-K′ shows a close correlation of the maximum with the western edge of the Hafod-y-Fedw igneous complex (p. 56). This consists of microdiorite and quartz-microdiorite intruded into a succession of mudstones with thin sills of dolerite and microtonalite. Detailed examination reveals that only some of the intrusions are associated with significant magnetic anomalies and, as Rice and Sharp (1976) have observed, 'the magnetics over them was variable and showed no particular trend'. Field measurements of susceptibility by Allen and others (1979) showed values up to 3×10^{-3} SI units from some of the thin sills and on cleavage and joint faces in the mudstones. The rocks in the Hafod-y-Fedw complex appear to have very low susceptibilities. Remanent magnetisation in samples of the mudstone and minor intrusions was negligible. The main magnetic mineral present is pyrrhotite. The conclusion is that the magnetic anomalies, despite their apparent correlation with the intrusions, are caused by metasomatic pyrrhotite, disseminated through suitable host rocks such as mudstones of the Clogau and Maentwrog formations, perhaps co-genetic with the quartz veins of the Dolgellau Gold-belt.

SEISMIC SURVEY

As a result of Powell's (1956) prediction that low density sediments in Tremadoc Bay caused the gravity gradient in the coastal strip, seismic refraction investigations were carried out (Griffiths and others, 1961; Blundell and others, 1964), and revealed the existence of four layers of characteristic velocities. These layers are not necessarily to be compared with unique lithological units but have a sufficiently distinct bulk velocity to produce refracted seismic waves with characteristic velocities, which may be identified as first—or more rarely as later—arrivals at the surface. The nature of the layers was a matter of debate, until the stratigraphy was clarified by the drilling of the Mochras Farm Borehole, allowing velocities to be measured in the formations penetrated. Bullerwell and McQuillin (1969) reported the results of a Geological Survey seismic reflection survey in Tremadoc Bay in which reflectors which correlate with the four layers may be seen. An interpretation of a profile which approaches the Mochras Farm Borehole is shown in Figure 24.

The uppermost of the four layers has a velocity of 1.5 km s^{-1} and reaches a maximum depth of 80 m below sea level. Blundell and others (1964) consider that it (their layer 1a) is not present everywhere in Tremadoc Bay, but fills overdeepened glacial valleys. Sparker records (Blundell and others, 1971) show that the layer has irregular interfaces, with scattering due to wash-outs. The layer is considered to be formed of Quaternary deposits, which are penetrated in the Mochras Farm Borehole to a depth of 77 m and which have an average velocity determined from the sonic log of 1.9 km s^{-1}.

The second layer (layer 1b of Blundell and others, 1971)

shows horizontal stratification and overlies a slight angular unconformity seen on their sparker records. It has an average velocity which varies between 1.8 and 2.4 km s^{-1}, and reaches a maximum depth of 600 m near Morfa Dyffryn, thinning to the west. The base of the layer may be represented by a strong reflection at a maximum two-way travel-time of 0.5 s on the Geological Survey reflection profile 2B (Bullerwell and McQuillin, 1969). This layer is correlated with the Tertiary section in the borehole; the average velocity determined from the sonic log is 2.4 km s^{-1}, with a variation between 1.86 and 3.02 km s^{-1} averaged over 50 m sections (Masson Smith, 1971); this is closely comparable with the results of the refraction work.

The third layer has velocities varying between 3.0 and 4.0 km s^{-1}, with the higher values predominating. It may reach a depth of about 2.5 km beneath Morfa Dyffryn (although depth estimates based on the refraction data are tentative being based on unreversed parts of profiles). It thins to the west, so that it has wedged out before the Cambrian outcrop on the Lleyn. This layer corresponds to the Jurassic and Triassic rocks intersected in the borehole. Sonic velocity data are available for about 23 per cent of the Jurassic section and give an average velocity of 4.26 km s^{-1}, varying between 3.87 and 4.77 km s^{-1} (Masson Smith, 1971); no data are available for the Triassic section. Velocity measurements on Triassic rocks, collated by Blundell and others (1964), are grouped about 3.0 km s^{-1}, but estimates from boreholes in Cheshire by Smith and Burley (1979) average about 4.2 km s^{-1}; the value of these estimates may not be great in view of the distinctive Triassic lithofacies intersected in the borehole. The base of the layer may correlate with an eastward dipping event on the Geological Survey reflection profile 2B, which reaches a maximum two-way travel-time of 1.5 s some 4 km W of Mochras Farm Borehole, and which accords with results from the refraction method. There is a strong reflection at 1.0 s two-way travel-time, which may mark the base of the Jurassic rocks (extrapolating estimated depth to the borehole); beneath is a more transparent zone, presumably comprising the Triassic and (if it is present) the Permian section.

The layer beneath is the lowest recognised; it may have a velocity of 5.5 km s^{-1} determined from refraction shooting but shows no distinctive structures on the reflection profile. Velocities in the Cambrian rocks range from about 4.0 to over 6.0 km s^{-1} (Appendix 2), so that the velocity for the layer falls into this range or that for the Ordovician (Blundell and others, 1964). The upper surface is presumed to be the unconformity between the Mesozoic and Palaeozoic rocks, and velocities of the head wave might be expected to be variable because of structures truncated by the erosion surface.

In addition to investigations of the layering, Griffiths and others (1961) determined the position of an interface between the low velocity material and the Cambrian rocks to the east at Morfa Dyffryn. The results are discussed in Chapter 5 under the section on the Mochras Fault.

ELECTROMAGNETIC SURVEY

During the course of the Mineral Reconnaissance Programme sponsored by the Department of Industry, airborne electromagnetic (EM) data were obtained for the eastern part of the Harlech Dome on the same flights as the magnetic data described above. The Scintrex HEM co-axial system, which operates at a frequency of 1600 Hz, was used and maintained at a terrain clearance of about 30 m. The resulting maps at a scale of 1:10 560 show the amplitude of the In-phase component and ratio of In-phase to Out-of-phase components; they are available from offices of the Geological Survey. The original flight records are held by the Regional Geophysics Research Group at the Geological Survey. A typical 10 km square is reproduced in Figure 32. The results of the airborne and consequent ground surveys have been described in Allen and others (1979). Some conclusions are reviewed here.

The main causes of the anomalies are various conductive beds. A discontinuous series of EM anomalies of at least 300 ppm coincide closely with the southern and northern parts of the outcrop of the Clogau Formation, part of which is shown on Figure 32. In the central part (between grid lines 323 km N and 330 km N), anomalies are weak or absent. In general the anomalies are asymmetrical in cross-section and elongated in plan. Ground follow-up of these anomalies confirmed their position within the limits of accuracy of position fixing, and their correlation with the outcrop of the Clogau Formation.

The asymmetry of the gradients on either flank of the anomalies is indicative of the dip of the causative body; in general an eastward dip is indicated, which might be expected from the structural evidence. Detailed sampling and subsequent laboratory studies identified the conductor as a combination of stringers of carbonaceous material and a network of sulphides (including pyrrhotite) deposited along bedding and on cleavage and fracture planes in a dark grey mudstone. Groundwater may make an additional contribution by healing breaks of continuity in the network. It has been contended by Allen and others (1979) that some of the sulphides are co-genetic with the quartz veins of the Dolgellau Gold-belt, and thus EM could be used as a pointer to potentially mineralised areas if suitable host rocks are known to be present.

Along the eastern margin of the district a broken line of anomalies up to 400 ppm In-phase component is coincident with the outcrop of the Dolgellau Member of the Cwmhesgen Formation, also shown on Figure 32. On the EM map it is possible to identify the position of the outcrop and of faults which displace it, including where it is under drift [c. 80 26] or beneath a thin veneer of Rhobell Volcanic rocks [77 27]. Ground traverses confirmed the position and correlation of the anomalies.

A traverse across the formation at Ogof Ddu (near Criccieth) to the north-west of the Harlech district identified sites for collection of the conductive rock. Laboratory study showed the presence of amorphous carbon in samples from the 'Black Band' and from elsewhere in the formation, corresponding to low resistivity values; this gives clear evidence of the cause of the anomaly. Similar stratigraphic conductors were identified in the Maentwrog Formation [71 20 and 74 28] and Aran Volcanic Group [79 20 and 80 23], but no studies of the petrology were carried out in these cases.

In addition to the stratigraphic conductors, other causes of anomalies have been identified. Anomalies due to conductive overburden were commonly encountered during the ground survey over saturated clay or peat-filled hollows. The effect may be distinguished by a high ratio of In-phase to Out-of-phase components. A strong anomaly coincident with the Mawddach estuary [65 18 to 70 19] is due to sea-water or to alluvium saturated with sea-water; the water has a high conductivity due to the dissolved salt.

It is important to recognise anomalies caused by artifacts such as power-lines, water-pipes or radio transmitters, examples of which may be seen on the data for the district. Additionally anomalies may be spurious, or may not be identifiable on the ground; Allen and others (1979) give a list of such airborne anomalies in the area. IFS

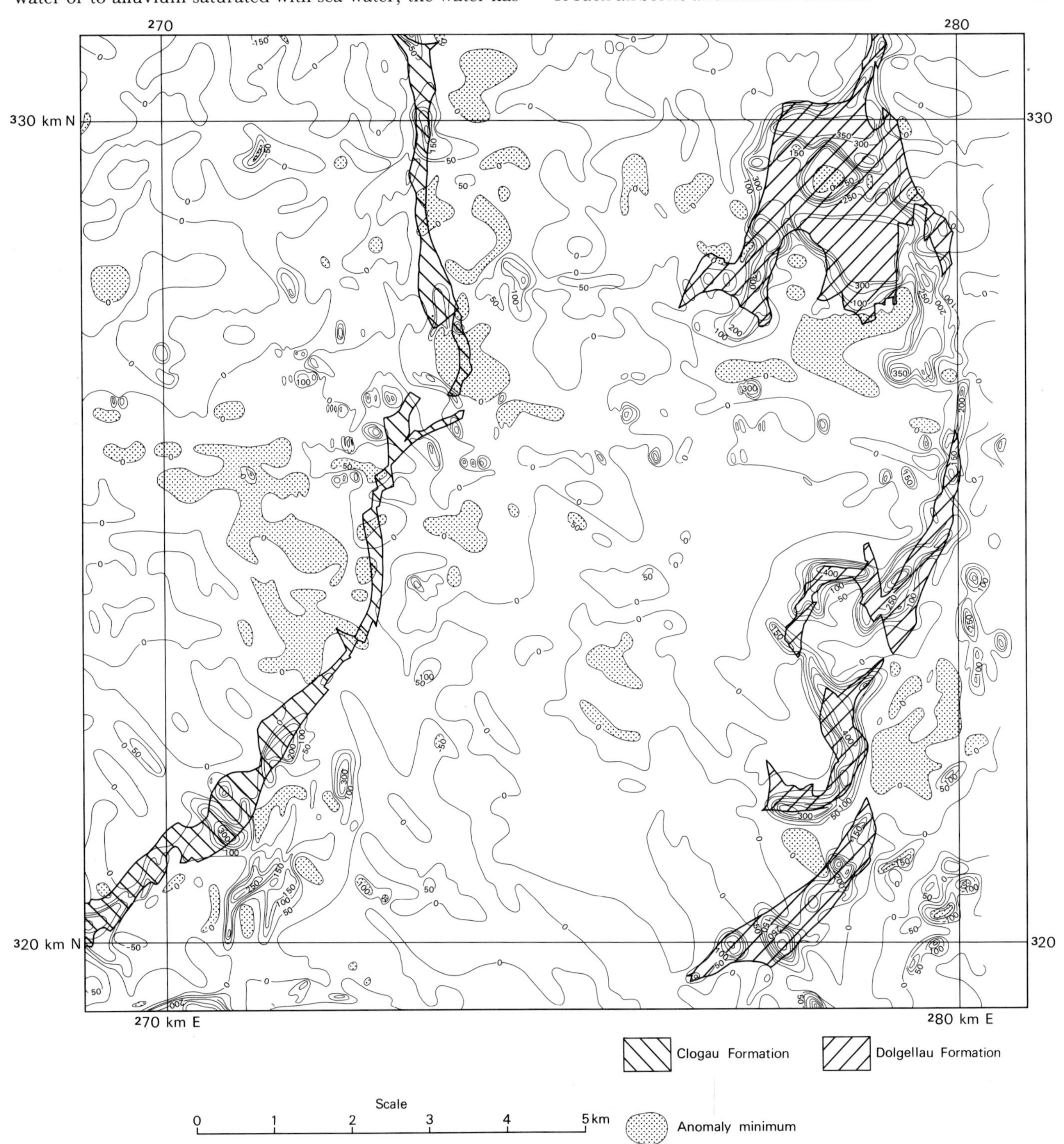

Figure 32 Airborne electromagnetic in-phase component map of grid square SH 72 using Scintrex co-axial system operating at 1600 Hz. Contour interval is 50 ppm. Also shown are the areas of outcrop of Clogau Formation and the Dolgellau Member of the Cwmhesgen Formation

CHAPTER 7

Economic geology

The Harlech Dome has long been known to offer a wide variety of useful raw materials, and it is suspected that base metals were mined in the area in antiquity (Morrison, 1975). Copper, lead and silver were certainly mined in the 18th century. During the 19th and early 20th centuries gold, copper and manganese were profitably mined, and the Dolgellau Gold-belt (Andrew, 1910), from which most of the gold was obtained, lies entirely within the district. Mining effectively ended when the last manganese was extracted in 1928. In 1968–73 Riofinex discovered, but did not exploit, the first porphyry copper deposit in Great Britain. In a small way slate has been successfully exploited, whereas sand, gravel, building stone and aggregate, though not extracted on a large scale, are all readily available.

SYN-SEDIMENTARY METALLIFEROUS DEPOSITS

The only bedded, syn-sedimentary metalliferous deposit to have been mined in this area is manganese, extracted from the Hafotty Formation (Figure 33). The black mudstone in the Dolgellau Member has been found to contain low levels of uranium (Ponsford, 1955). Beds of oolitic ironstone within the Aran Volcanic Group are thin and of no economic interest. However, indications have been discovered by Allen and others (1979) of a metallogenic horizon within the Aran Volcanic Group at the base of the Benglog Volcanic Formation. Geochemical and geophysical surveys carried out around Cae'r-defaid [795 235] have revealed anomalous concentrations of base metals in rocks, soils and stream sediments at this level (Cooper and others, 1983).

Manganese

Manganese ore forms a bedded deposit of rhodocrosite, spessartine and rare rhodonite within the Hafotty Formation (see p. 10). It weathers to todorokite (Glasby, 1974), and early mining activity was concerned mainly with removing the secondary products. Dewey and Bromehead (1915) and Dewey and Dines (1923) did not recognise spessartine in the ore. They, quoting an unpublished account by Goodchild in 1893, believed that there were two ore horizons. Woodland (1939) showed that the lower of Goodchild's two horizons is the principal ore-bed. It occurs constantly about 9 m above the base of the Hafotty Formation, but has not been traced farther north than Moel Ysgyfarnagod (Figure 33) nor north of Afon Gamlan. The bed is usually 30 to 45 cm thick, but Dewey and Dines (1923) record thicknesses of 0.9 m on average in the Egryn Mine [619 215] near Barmouth. At Cae-mab-seifion [691 219], on the easternmost limits of the area containing the ore-bed, it has thinned to 20 to 25 cm.

Thin beds, lenses and nodules of spessartine-quartz rock occur throughout the Gamlan Formation; the ore locality [6114 3088] near Ffridd-llwyn-Gurfal, listed by Dewey and Dines (1923), is within that formation. This probably corresponds with the higher of Goodchild's two ore horizons, the Upper Manganese Bed, which he put in the Menevian. Compared with the lower ore-bed this one is thin and impersistent.

Mining for manganese first took place in this area in the middle 19th century when only the secondary oxides of the weathered zone were extracted. From 1892 to 1928, with a short break between 1909–1913 during which there was no production, the primary ore was mined. During this period about 44 000 tons of ore were extracted (Thomas, 1961). In 1957 a mining company took out several leases in the area north of Barmouth, but there were no subsequent developments, though there is a considerable resource of manganese-rich material left in the area.

The ore, which dips at angles up to 70°, was extracted mainly from open workings, which extend for several kilometres along the outcrop, though in places levels and adits were driven into and along the bed. Little attempt was made to follow ore beneath drift.

Dewey and Dines (1923) gave grades varying between 27 to 34 per cent Mn for the worked ore, but do not specify whether they are for fresh ore, enriched weathered ore, or mixed. These values fall below those currently used to define ore by the U.S. Bureau of Mines (Weiss, 1977). In six analyses of the fresh, chocolate-red or yellow ore given by Woodland (1939) the Mn content calculated from MnO values ranges from 28.5 to 31.75 per cent. The silica content is high, ranging from 21.06 to 27.17 per cent; up to 13.6 per cent of this is present as free silica. Rhodocrosite and spessartine are the principal ore minerals, the garnet ranging from 42 to 55 per cent of the total of the two. The less common bluish black bands are substantially richer in Mn with values up to 39 per cent. In addition to minor finely divided todorokite, they contain rhodocrosite and spessartine in the ratio of about 3:1. An analysis of a sample of a whole band taken from the Gamlan Formation by Dr I. Basham of the Geological Survey gave only 10.8 per cent Mn.

Uranium

Ponsford (1955) measured between 0.0014 and 0.006 per cent U_3O_8 in samples from the Dolgellau Member in areas around the Harlech Dome. These levels are too low to be of economic interest. The highest contents are in light grey silty mudstone bands within the black mudstone, and the highest value of all came from the light grey bed at the base of the 'Lower Dolgellau Beds' at Ogof Ddu which, though outside the district, has been examined by the authors and considered to lie within the Ffestiniog Flags Formation.

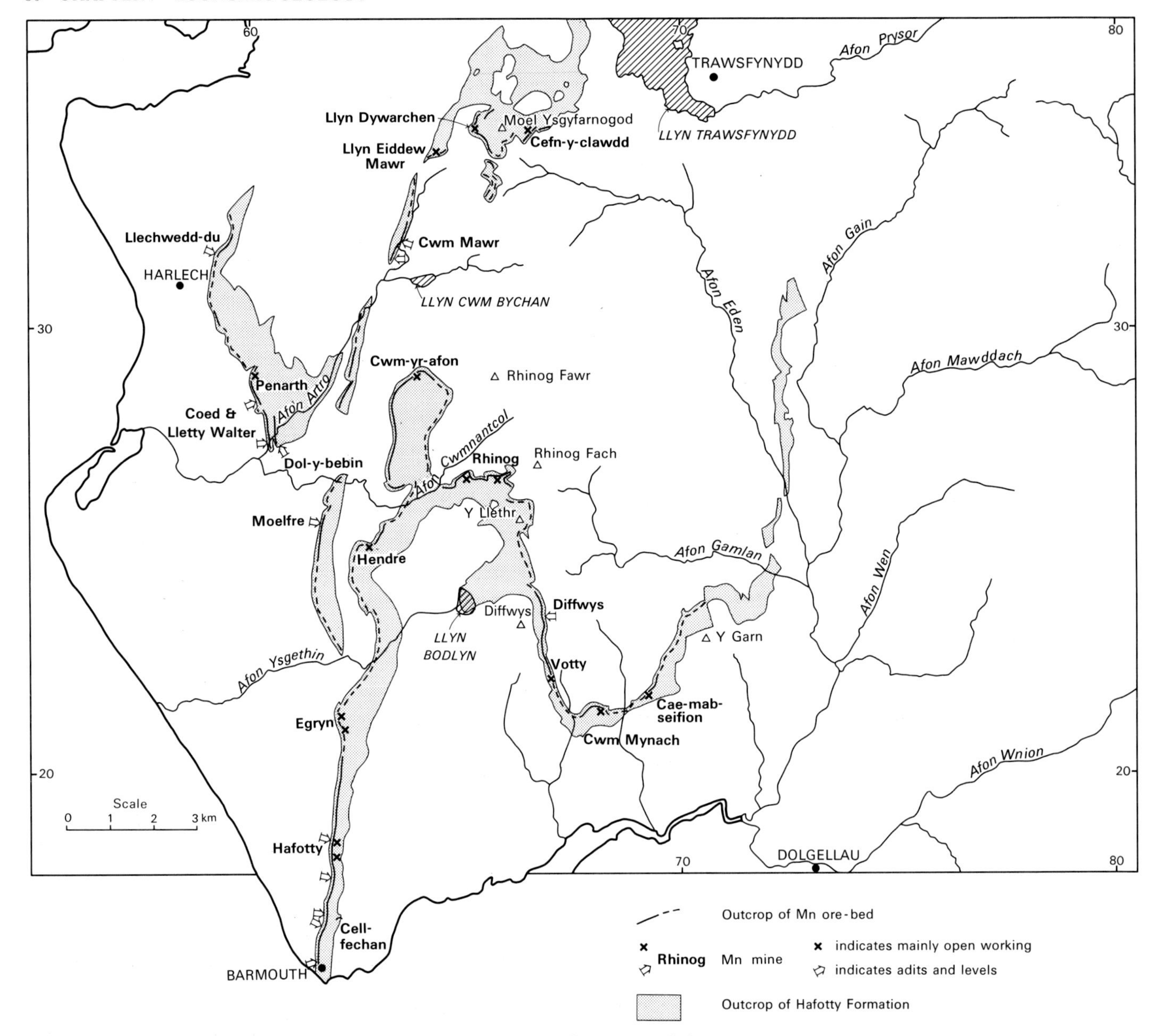

Figure 33 The principal manganese mines and the area of outcrop of the Hafotty Formation

PORPHYRY-COPPER MINERALISATION

The existence of porphyry-style mineralisation was unknown in Great Britain before the Coed y Brenin orebody was discovered and evaluated by Riofinex in 1968–73 (Rice and Sharp, 1976), though reference to disseminated copper minerals in the country rock hereabouts had been made by Ramsay (1866) and Hunt (1887). The mineralised body is spatially associated with the Afon Wen microtonalite intrusion complex, which is emplaced within the Ffestiniog Flags Formation (p. 55) and thought to be comagmatic with the late Tremadoc Rhobell Volcanic Group. PMA

K-Ar age determinations on mica collected from borehole core[1] within the mineralised zone at Coed y Brenin fall in the range 370 to 417 Ma (Table 10). The data demonstrate two distinct age populations. Since it is most unlikely that the close cluster of ages represents the same degree of partial argon loss from all the samples, it is suggested that the mean age for each population reflects a real geological event, such as increased temperature and/or stress, which has reset the K-Ar clocks. The mean age of the elder group (409 ± 7 Ma; i.e. twice the standard error) is not significantly different from the mean age determined for micas separated from gold-quartz veins (Table 11) and presumably records the same event. The younger group, however, has given a significantly lower mean age of 374 ± 5 Ma; it is not clear to what particular event this age may be related. CCR

The older of the two mean ages does not necessarily date the mineralisation. Geochemical evidence provided by Allen and others (1976) suggests that the mineralisation probably relates to the Rhobell Volcanic Group magmatism in the late Tremadoc.

[1] with permission of Riofinex Ltd

Table 10 K/Ar age determinations on micas from Coed y Brenin

Sample No*	Borehole No	Depth in ft	Material	%K[1]	nl/g Rg^{40}Ar	Age in M.a. & error (2-sigma)	Mean ages in M.a.
NW-75/52	CB 85	52	Mica adjacent to thick quartz vein	8.642	139.45	373 ± 10	
NW-75/66	CB 62	430	mica from sheared 'diorite'	3.022	48.334	370 ± 10	375 ± 5
NW-75/77	CB 50	610	mica from sheared mixed 'diorite' and sedimentary rock	4.163	68.336	379 ± 12	
NW-75/78	CB 62	565	mica from 'diorite'	3.142	55.519	405 ± 12	
NW-75/79	CB 61	182	mica from 'diorite'	4.000	72.457	414 ± 13	
							409 ± 5
NW-75/80	CB 61	146	mica from 'diorite'	3.520	61.618	401 ± 12	
NW-75/81	CB 61	473	mica from 'diorite'	3.510	64.094	416 ± 13	

* All data are held in Mineral Sciences and Isotope Geology Research Group file under Lab. No. 1198.
[1] Low K contents of some samples represent incomplete mineral separation and dilution mainly by quartz, rather than altered mica.

Investigations reported by Allen and others (1979) show that within the Afon Wen Complex there are indications of extensions of the Coed y Brenin mineral zone, or even of separate orebodies. For example 1.5 km to the south of it at Bryn Coch [740 245] 19 randomly collected surface rock samples contain up to 0.67 per cent Cu, with a mean value of 1175 ppm Cu. Soils collected over the probable northern extension of the Afon Wen intrusion complex at Hafod Fraith [749 275] are copper rich, and soils over the discordant feeder to the Afon Wen complex carry as much as 3500 ppm Cu. Copper minerals, both in veinlets and disseminated, have been found in the southernmost parts of the complex in quartz microdiorite near Pont Wen [736 226].

Allen and others (1976) showed that the microtonalite is unusually copper rich in all parts of this area. Their investigations around the Range and Dôl-haidd intrusions (p. 54) showed rare spots of malachite and some chalcopyrite disseminated in the microtonalite and on joints. The most heavily mineralised parts of the Dôl-haidd intrusion are in Nant Braich-y-ceunant near a trial working [7675 3614]. Here the intrusion has undergone propylitic alteration and there is abundant pyrite and chalcopyrite, both disseminated and with calcite in veinlets in the country rock and minor offshoots from the main intrusion. One sample of microtonalite from here contains 0.27 per cent Cu.

In addition to the disseminated mineralisation, Allen and Easterbrook (1978) showed that the copper mined at Glasdir until 1914 was taken from a breccia pipe which probably formed during the late Tremadoc magmatism. Five more pipes have been located (p. 59) and traces of chalcopyrite identified in one of them.

Coed y Brenin porphyry copper deposit

This deposit [749 257] is located mostly under thick boulder clay in a topographic depression near Capel Hermon on the west of the Afon Wen. Rice and Sharp (1976) recognised three phases of intrusion within the deposit (p. 55) and commented that mineralisation is confined to rocks of the oldest phase, the so-called 'Older diorite' and the adjacent sedimentary rocks. The main intrusion of this phase shows a sharp contact on the west, but is brecciated and full of xenoliths of wallrock in the upper part on the east.

The deposit consists only of the hypogene zone; any zone of supergene enrichment that formed was presumably removed during the Pleistocene glaciations. Chalcopyrite occurs mainly in hairline fractures and as fine disseminations. Traces of molybdenite occur along joints, fractures and,

Table 11 K/Ar age determinations on samples from gold-quartz veins

Sample No	Mine	Grid reference	%K	nl/g Rg^{40}Ar	Age and error (Ma)
NW 75/6	Gwynfynydd	7354 2818	4.868	82.641	390 ± 12
NW 75/26	Blaen-y-Cwm	7131 2233	5.032	88.433	403 ± 12
NW 75/34	Clogau	6740 1998	4.645	83.326	410 ± 13
NW 75/41	Hafod Uchaf	6524 2096	7.008	121.60	410 ± 9
NW 75/41	Hafod Uchaf	6524 2096	7.746	143.14	
NW 75/42	Craigwen	6548 1886	7.021	125.37	408 ± 13
NW 75/45	Foel-ispri	7046 2011	3.560	64.746	406 ± 9
NW 75/45	Foel-ispri	7046 2011	6.166	106.35	
				Mean age	405 ± 6 Ma

locally, in quartz-calcite veinlets. There are two zones of more than 0.2 per cent Cu, within which both molybdenite and gold are present. The classic pattern of metal zoning is recognised with Cu–Mo in the centre, surrounded by an Fe-rich pyrite zone, with a zone peripheral to this of Pb–Zn, mostly in veins. The patterns of hydrothermal alteration zones is incomplete with only the phyllic and propylitic zones identified with certainty. It is possible that the sill-like shape of the host intrusion prevented the establishment within it of a convective system big enough, in comparison with the more common stock-like hosts, to create the higher temperature alteration zones. PMA

Shepherd and Allen (*in press*) found that five types of fluid inclusion characterise this deposit. The simplest and most abundant are two phase inclusions containing an aqueous liquid and vapour phase enriched in CO_2. More complex inclusions contain, in addition, a solid opaque mineral phase or cubic halite crystal. There are also aqueous inclusions distinguished by the presence of liquid CO_2 at room temperature. The quartz-sulphide and quartz-carbonate veins are characterised almost exclusively by simple aqueous or liquid CO_2 inclusions. By comparison with deposits in the USA (Nash, 1976), these show much lower temperatures and salinities, but are wholly consistent with their specific association with phyllic alteration. However, in quartz veinlets and along fractures in quartz phenocrysts in the host rocks the inclusions show a prominent development of halite-bearing types. These are more typical of those described by Nash and probably represent the original copper-bearing fluids which generated the disseminated and microfracture controlled Cu–Mo orebody before the initiation of vein mineralisation. TJS

Glasdir copper mine

This mine [741 223] is located on the south-eastern side of a prominent rocky knoll about 15 km W of Llanfachreth. It operated on a small scale in the middle of the 19th century, but increased output later. From 1872 until closure in 1914 over 13 000 tons (Table 12) of concentrate, containing about 10 per cent Cu and profitable quantities of gold and silver, left the mine for smelting (Morrison, 1975). Lead production is also recorded from Glasdir (Hunt, 1881), though the presence of lead minerals in the deposit has not been confirmed. The workings were opencast at first, but were later extended underground.

Andrew (1910) described the lode as shattered country rock impregnated with copper minerals, and commented on the unusual lack of quartz in the orebody. On re-examining the mine Allen and Easterbrook (1978) concluded that the orebody, shaped like a flattened inverted cone, lies within the margin of an intrusive breccia pipe emplaced within the Ffestiniog Flags Formation.

The ore, which never ran at more than 2 per cent Cu (Phillips, 1918), contains pyrite, marcasite, arsenopyrite, minor sphalerite and chalcopyrite, which in places is rimmed by covellite. The gangue includes calcite and dolomite in addition to the minerals present elsewhere in the breccia. Shepherd and Allen (*in press*) found that the fluid inclusions have characteristics in common which those in the porphyry copper deposit at Coed y Brenin. An analysis of the inclusion gases revealed a similar high CO_2 component (mean 0.7 mole % c.f. 0.6 mole % Coed y Brenin), thereby providing an important latent energy source to cause the brecciation of the country rocks.

Table 12 Output of copper from the principal mines

Mine	Year	Output of copper in tons of ore[1]
Old Clogau	1845	560
and Vigra	1846	34
Old Clogau	1860	43
Dolfrwynog	1844, 1847	67
Glasdir	1872–1914	13 077
Turf	1817	25
	1824–47	1 742

[1] For Glasdir the output is in tons of concentrate running at an average of 9 to 10 per cent. Copper contents for ore from the other mines is not known.

VEIN MINERALISATION

Quartz-sulphide veins, emplaced in Cambrian and Ordovician rocks after the Caledonian folding and metamorphism, have been worked in the district for gold, copper, lead, zinc and silver. They intersect most formations from the Rhinog Formation upwards, though the majority are within the Clogau Formation.

The veins commonly form anastomosing systems in places along faults. Some extend for 3 km or more, and they vary in thickness along their lengths. Individual veins have been recorded in excess of 6 m thick. Most are vertical, though dips as low as 45° are known. Veins with all orientations are known, but they commonly trend between 030° to 070°. There is a subsidiary complementary set trending 290° to 315°.

Mineralogically the veins in the Gold-belt have been classified by Huddart (1904) and Gilbey (1969). In the simplest terms they are characterised either by pyrite-chalcopyrite-pyrrhotite dominant or sphalerite-galena dominant assemblages. Less abundant minerals listed by Gilbey (1969) include arsenopyrite, marcasite, mackinawite, tetrahedrite, cobaltite and tellurbismuth, the latter claimed by Kingsbury (1965) to have previously been misidentified as tetradymite. Gilbey (1969), however, noted minor quantities of tetradymite associated with the tellurbismuth. He also lists a number of other rare minerals including gold tellurides. Native gold occurs either dispersed through the quartz or intimately associated with the sulphides. In certain instances the gold is present either as replacement or exsolution blebs in pyrite and arsenopyrite. Some of the sphalerite-galena dominant veins were worked for silver. Forbes (1867) recorded polytelite, native silver and richly argentiferous silver at Tyddyn Gwladys, and Hunt (1875) gives 50 to 60 oz Ag per ton of galena from this mine and 6 to 11 dwt Au. Analyses of gold from several mines given by Forbes (1867) show it to be a silver-rich gold/silver alloy. For gold at Clogau two analyses are 90.16 Au, 9.26 Ag and 89.93 Au and 9.24 per cent Ag. At Gwynfynydd it is 84.89 Au, 13.99 per cent Ag and at Cefn Coch 76.40 Au, 22.70 per cent Ag. Small amounts of epidote, chlorite, sericite, calcite and dolomite are present in the veins in addition to the ubiquitous fragments of chloritised mudstone. Supergene alteration, which is restricted in occurrence, has given rise at surface to malachite, azurite, lepidocrocite-goethite and erythrite.

According to Gilbey (1969) wall rock alteration is limited to within 6 m from the contact. In the peripheral parts, and increasing in intensity inwards, there are chloritisation and hematitie-goethite concentrations along fractures and joints. An inner iron-rich zone less than 1.6 m from the contact contains pyrite, arsenopyrite, sphalerite, galena and chalcopyrite, in addition to sericitisation and quartz impregnation. In places the immediate wall rock is silicified. Allen and others (1979) record epigenetic pyrrhotite developed extensively in cleavage planes, joints and veinlets in areas around the mineralised quartz veins. PMA

Age of vein mineralisation

Moorbath (1962) obtained a model lead age with a mean of 430 Ma on minerals from this area. K/Ar determinations carried out by Ineson and Mitchell (1975) on samples of altered wall rock from two mines in the area gave ages of 368 ± 5 Ma to 397 ± 5 Ma. New K/Ar ages have been determined on micas from six mines in the area. All the samples are composites made up from material collected on dumps (Table 11).

All ages have been calculated using the decay and other constants recommended in Steiger and Jager (1977). The errors quoted are at the two sigma level and represent estimates of analytical uncertainty only. No individual age differs from the mean by more than twice the standard deviation, derived from the analytical errors, and hence the spread of values can be attributed to random analytical error. The mean value of 405 ± 6 Ma may be taken to be the best estimate of the age of mineralisation.

This early Devonian mean age is in accord with field evidence that shows the veins to post-date folding and metamorphism, and all six mines sampled lie within the Gold-belt. There are, however, veins (e.g. at Dôl-frwynog) mined for copper around the Coed y Brenin copper deposit that may be genetically related to the late Tremadoc mineralisation. CCR

Metal zoning in the veins

There was a strong belief among early miners that gold did not occur in veins below the level of the Clogau Formation. Gilbey (1969), however, records rare gold in veins within the Gamlan Formation, and comments that at the Clogau Mine [675 199], the most productive in the area, gold was retrieved from below the black mudstone formation. Andrew (1910), who visited this mine while it was working, claimed the contrary. A certain amount of stratigraphic control to the vein mineralisation, however, is likely, and was invoked by Andrew (1910) in his discussion of the genesis of the veins. Shepherd and Allen (*in press*) suggest that the presence in the Clogau Formation of free carbon, also invoked by Andrew (1910), and pyrite-pyrrhotite assemblages, both of which do not occur at stratigraphically lower levels, had a significant buffering effect on the fugacities of oxygen and sulphur, resulting in the destabilisation of gold complexes in solution and the preferential precipitation of gold at or above the Clogau Formation.

There has been a good deal of speculation about the optimal conditions necessary for the concentration of gold within a vein (Hall, 1975). The gold tended to occur in 'bunches' or pockets within the principal veins, and the control of the location of these pockets has been attributed to the presence of vein intersections, which juxtapose black mudstone in one wall with 'greenstone' in the other. Andrew (1910) recorded that the gold was always associated with bluish quartz and was concentrated adjacent to mudstone inclusions in the Clogau Mine. PMA

Fluid inclusions

Shepherd and Allen (*in press*) found that primary fluid inclusions in the quartz mainly contain simple two-phase systems comprising brine with a salinity of less than 6 wt. % NaCl equivalent and a co-existing vapour. The additional presence of solid, opaque daughter minerals or liquid CO_2 was rarely observed. Some inclusions contain high salinity brines (16 to 19 wt. % NaCl equivalent). The temperature-salinity pattern suggests an uncorrected temperature of formation in the range 135 to 350°C. Applying pressure corrections for loads not in excess of 2.5 kb (the assumed lithostatic load at the time of mineralisation), the maximum temperature would not exceed 430°C, and more likely would be about 350°C. These are considerably lower than 535°C at 2.4 kb predicted by Gilbey (1969). It was found that inclusions containing high salinity fluids and higher than usual CO_2 levels characterised veins with abundant galena and sphalerite but no gold. TJS

Metallogenesis

Prior to Gilbey, little research had been undertaken on the genesis of the veins system. Archer (1959) had suggested that the Gold-belt veins lay in the core of a zoned system covering all North Wales and derived from a single post-Carboniferous mineralisation phase. Gilbey (1969) believed that there was zonation within the Gold-belt, the higher temperature and first-formed veins being in the west. He described the following ideal order of deposition for the veins:

1 Deformation along fissures with accompanying brecciation and gouging, and the deposition of pyrite and arsenopyrite in the wall rock.
2 Crystallisation of quartz and calcite in open fissures, with accompanying cobaltite in the west and arsenopyrite-pyrite in the east.
3 Chalcopyrite/pyrrhotite and lesser-marcasite replaces gangue.
4 New spaces occupied by galena, sphalerite and pyrite with second generation chalcopyrite, quartz and calcite (± dolomite and siderite).
5 Crustification assemblages of calcite, marcasite, sphalerite with rare galena and arsenopyrite.

Gilbey (1969) regarded the veins at Cae Mawr to represent a distinct mineralisation in which the ore minerals were derived from the 'greenstone' country rock (the Hafod-y-fedw intrusion, p. 56). For the other veins he postulated an origin by mixing of magmatic and juvenile fluids from a subjacent batholith with connate water, and the penetration of the resultant fluid through an open fracture system. By contrast Shepherd and Allen (*in press*) prefer to generate the hydrothermal fluids by dewatering of lowermost Cambrian and Precambrian rocks through hydraulic fracturing during end-Caledonian uplift.

History of mining

The announcement of the discovery of gold at Cwm Hesian mine in 1844 came at a time when mining in this area had been going on a least since the late 17th century and possibly since ancient times (Morrison, 1975). The veins were then primarily being worked for copper, lead and silver, but because records are so poor the production is largely unknown.

According to Morrison (1975) interest was expressed in the veins on Foel Ispri as early as 1760, and in the period up to 1853, when gold was discovered there, the mines (known at different times as Foel Ispri, Prince of Wales, Imperial, Voel, Hafod-y-morpha, etc) primarily produced lead and silver (Table 13). There are records of zinc production from there in the years after 1856. Lewis (1967) claims that the

Table 13 Production of Pb, Ag, Zn from the principal mines

Mine	Year	Production		
		Lead (ton)	Silver (oz)	Zinc ore (ton)
Tyddyn Gwladys	1848, 1851	21.8		18.9
Prince of Wales	1856–57, (1868)	125.75	110	(20)
Imperial	1868 & 1870	70	382	20
Foel Ispri	1890–91	0.25	14	130
Caegwernog	1874			19
Harlech Lletty (Llanfihangel-y-Traethau)	1875	3.5		
Pioneer	1881	28		
Dolgelly	1849	11		
Glasdir	1876, 1879	7.75		

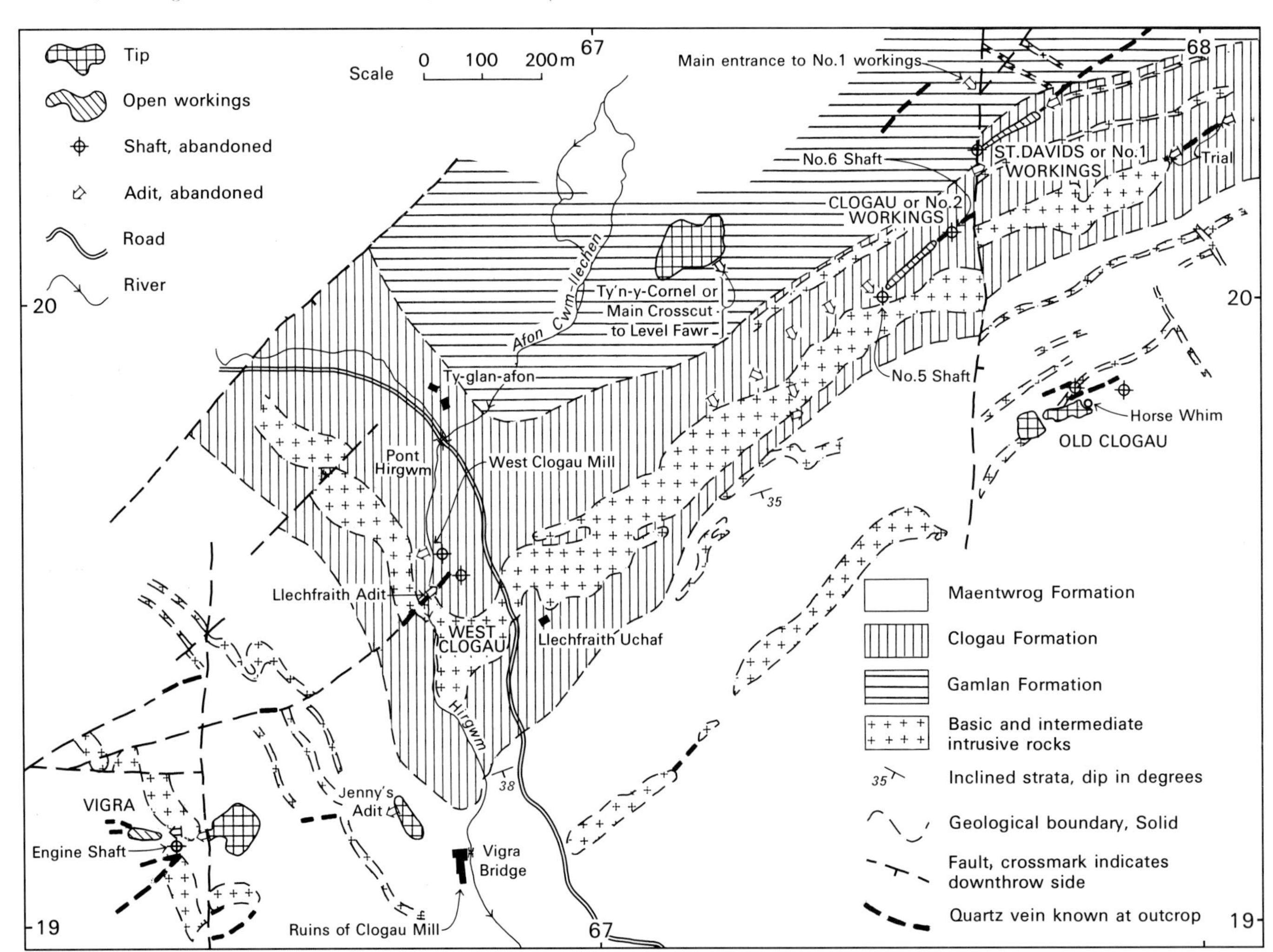

Figure 34 Geology of the area around the Clogau and Vigra mines

Cefn Coch mine was producing lead in the 18th century. Both of the Cwm Hesian mines [742 282, 737 277] and Tyddyn Gwladys [734 268] were worked as lead-silver mines in the early 19th century, and some authorities (see Hall, 1975; Morrison, 1975) believe that Gwynfynydd was also in operation for lead at that time. Tyddyn Gwladys was one of the richest silver mines in the area, and it is claimed (Hunt, 1875) that in 1865 a string of silver ore there yielded 2200 oz Ag per ton.

The Old Clogau [679 198] and Vigra [663 191] mines (Figure 34) were both opened for copper and in operation before 1844. They continued to produce copper (Table 12) when gold became the main object of mining activity in the area. In the area around Dôl-frwynog [747 257], which Morrison (1975) believed was mined in ancient times, there are over thirty old shafts and adits. They continued producing copper until 1874, in the later years with some gold.

Claims for the first discovery of gold in this area came in 1844, and during the next 70 years nearly every quartz vein in the area was tried. Gold production was sporadic, but in total over 130 000 oz were retrieved. Several mines produced a small amount of gold in 1853–54, and again in the early 1860s when production from Clogau dominated the output from about 20 mines. However, it was not until after 1887, when Gwynfynydd opened and Clogau resumed production, that the gold-belt showed its worth. Output culminated in the period 1900–1905. At one time there were nearly 30 mines in operation but this number declined towards 1916, after which activity has been insignificant. Attempts were made to reopen Gwynfynydd, Clogau, Bedd-coedwr [742 286] and Graigwen in the 1930s, and in 1934–35 Bedd-coedwr mine produced the gold for Princess Marina's wedding ring. Both Gwynfynydd and Clogau have recently reopened and are currently (1985) working and being reassessed.

Production figures taken from various sources, and quoted by Hall (1975) and Morrison (1975) show immensely variable yields of gold per ton among the mines. The overall yield from the gold-belt is in the order of 0.5 oz/ton of crushed quartz. Peak output was 19 655 oz gold from 23 203 tons of quartz in 1904; that is 0.85 oz/ton.

Descriptive accounts of the gold mines have been furnished by Andrew (1910), Dewey and Smith (1922), and Dewey and Eastwood (1925). Hall (1975) and Morrison (1975), both of whom approached the matter historically, gave details of mine ownership. Morgan-Rees (1969) described the mining methods and equipment.

Details of mines

Clogau Mine

Of the several veins that have been worked or mined on Mynydd Clogau, the principal one is the St David's Lode. The vein is nearly vertical and trends NE, roughly parallel to the strike. It is offset by the Bryntirion fault, but according to Andrew (1910), can be traced from Vigra to Cae-mab-seifron, a distance of about 5.6 km. The workings of the Clogau mine are entirely within the St David's Lode, the most productive in the gold-belt. Within the mine it has a number of branches including the John Hughes and Paraffin or South Reef Lode. The vein cuts the Clogau and Gamlan formations and 'greenstone' sills (Figure 34). The main sulphides in it are pyrrhotite, chalcopyrite and pyrite. Arsenopyrite is common adjacent to shale inclusions (Gilbey 1969). A large number of other minerals, including gold tellurides, have been recorded in the veins by Andrew (1910) and Gilbey (1969).

Prior to work beginning on this lode in 1860, when the mine was under the ownership of T. A. Readwin, copper had been worked in Vigra mine, where one of the lodes may be a south-west extension of it, and at Old Clogau in a separate but parallel vein. All of these mines were under common ownership for much of their recorded history up to 1879, and they are not differentiated in the production figures.

Workings are extensive and said to include 19 miles of tunnels. The main level, Level Fawr [6720 2005] follows the vein for over 1 km. Except for two breaks totalling about 12 years the mine was in production from 1860 to 1911. In its heyday from 1900–1908, when up to 253 men were employed, no less than 12 000 tons of quartz were crushed annually; the peak production of 18 417 oz gold was in 1904. In all, over 81 000 oz of gold were won from this mine. Evidence furnished by Hall (1975) suggests that copper was never successfully extracted. Records show that attempts made to produce economically viable copper concentrates after 1898 included the installation of flotation plant, but failed.

Gwynfynydd Mine

Situated on the north bank of the Afon Mawddach about ½ km upstream from its confluence with the Afon Cain and surrounded by forest, this is the second largest goldmine in the area. Four veins—the Big, Chidlaw, Collett and James—trending between 060° and 080° with dips of 40° or more to the north, have been worked or tried on the west side of the Trawsfynydd Fault, and two—the New and Main—on the east side (Hall, 1975; Morrison, 1975).

On the west of the fault the veins intersect the Barmouth, Gamlan and Clogau formations and interstratified microdiorite sills, in some instances along faults. Bedding strikes roughly north and dips 30°–60°E. The longest and most important vein, the Chidlaw, can be traced for about 1 km and reaches 6 m in thickness. All seem to diminish westwards into a cluster of thin veins.

The veins on the east of the fault intersect the eastward-dipping Maentwrog Formation and microdiorite sills. The New and the Main veins, up to 7 m thick, merge eastwards and with some lesser parallel veins continue east-north-eastwards into the workings of the Bedd-Coedwr and Cwm Hesian East mines.

In contrast to the St David's Lode, the sulphide assemblages in all the veins at Gwynfynydd are pyrite-sphalerite-galena dominant with only minor quantities of chalcopyrite. Additional minerals recorded by Gilbey (1969) include arsenopyrite, marcasite, rare tetrahedrite and tellurides. Andrew (1910) recorded orpiment, pyromorphite and mimetite. Dewey and Smith (1922), quoting the mine manager, recorded that gold was distributed through quartz in high parts of the veins and was intimately associated with sulphides at lower levels.

Veins on the west of the Trawsfynydd Fault were exploited first under the ownership of T. A. Readwin in 1862, but production was insignificant until 1887, after which the mine was in operation with only two years of no production until 1916. Production was sporadic and yields varied from 2.3 oz per ton to 0.09 oz per ton crushed quartz in the most productive period from 1887 to 1904. The best year, according to Hall (1975), was 1888 when 8745 oz gold were extracted from 3844 tons of quartz taken from the Chidlaw Lode.

Work on the east side of the fault began in 1892 with the driving of No. 6 Adit [7373 2807] (Plate 12), and yielded good ore in the Main Lode in 1894. From then until closure, work was mainly carried on in this and the New Lode. In all, over 41 000 oz gold were produced from Gwynfynydd after 1862. Under the ownership of

Plate 12
Headframe of flooded inclines (bottom right) in No. 6 level at Gwynfynydd (L 1316).

Hillside Mining Company much development was carried out in the 1930s but there was no production. The mine reopened in the early 1980's and is currently working. No production figures are yet available.

TURF COPPER

The Turf copper mine [741 256] worked a unique deposit of copper-impregnated peat in the first half of the 19th century. The peat was no more than 60 cm thick and lay in a topographic depression less than ½ km W of the Coed y Brenin porphyry copper deposit, the presumed source of the copper. Henwood (1856) described the mine. About 70 acres of peat were dug and burned, the ash being sent to Swansea for smelting. The lower part of the peat layer was particularly rich in copper, and was often sent off unburned after drying. Apparently, peat which yielded less than 2.3 per cent Cu in the ashes was not dug. The copper was present native and in a variety of secondary salts. The site is now stripped of peat, but precipitation of copper minerals continues in the boulder clay which underlies it. Morrison (1975) and Archer (1959) give a total of 1767 tons of ore (see Table 12) from this mine, but other sources (Hall, 1975) quoting reports of extremely high profits for years as early as 1810, imply far greater yields.

SAND AND GRAVEL

The large resource of dune sand along the coast has not been utilised except on a limited scale about 1 km S of Mochras [558 258]. Though borings show thick beds of alluvial gravel under the Mawddach estuary and under the lower part of the Wnion valley, there is little gravel at surface apart from local alluvial terraces and scattered mounds of glacial deposits. Two of the latter appear to have been long dug on a minor scale: one [704 334] lies south of Trawsfynydd, the other is on the Welshpool road [*760 178*] east of Dolgellau.

AGGREGATE

There is a large number of small disused quarries, worked to supply local needs for aggregate, in all parts of the district. However, in 1976, only two were in production. They are on the Welshpool road [7397 1788], about 1 km E of Dolgellau, and in Abergwynant Woods [675 179], 2 km W of Penmaenpool. At both dolerite sills within the Ffestiniog Flags Formation are being quarried. A third quarry, near Trawsfynydd station [7135 3625], at which a greenstone sill within the Maentwrog Formation is extracted, operates spasmodically. The resource of stone for most purposes is enormous.

SLATE

Slate has been quarried from the Llanbedr, Ffestiniog Flags and Cwmhesgen formations, though with variable success. Slate has been extracted most successfully from the Llanbedr Formation. There are many disused quarries along strike on the western side of the Harlech Dome. The largest open quarry here [590 267] is at Llanbedr. Three kilometres north, at Llanfair [580 290], the slate was largely mined. Both here and south of Dyffryn Ardudwy at the disused Egryn

Quarry [605 206] and in a smaller one 1.5 km farther south [603 191], the slate was taken from just below the base of the Rhinog Formation. Purple and red slate from this formation was also extensively quarried at Cefn Cam [680 256] and to a lesser extent south-west of Llyn Trawsfynydd [c. 683 335], both in the central part of the dome.

Slate is known to have been extracted from the Ffestiniog Flags Formation only at Ffridd-y-Castell [783 234] near Cae'r-defaid, where most of the workings are underground. In this general area there are numerous trials within the Cwmhesgen Formation, particularly on Moel Cae'r-defaid [795 245]. Slate in this formation is black when fresh but the pyrite in it quickly weathers to an undesirable rusty brown which, according to Thomas (1961), has limited the development of quarries in this formation.

BUILDING STONE

Stone for building is not presently being worked. In the past rocks from within the district, including crystal tuff from the Benglog Formation and greywacke sandstones from parts of the Harlech Grits Group, have been popular. PMA

CHAPTER 8

Mesozoic and Tertiary rocks

THE MOCHRAS FARM BOREHOLE

The drilling of the Mochras Farm Borehole in 1967–69 established the age of much of the sequence present under Tremadoc Bay. This had been previously recognised as being of younger rocks than those lying onshore to the east, by its lower velocity, density and magnetic characters (Powell 1956; Griffiths, King and Wilson, 1961; Blundell, King and Wilson, 1964). The borehole, situated as far west as practicable on Morfa Dyffryn [5533 2594] was sunk to a depth of 1938.83 m, and proved substantial thicknesses of both Tertiary and early Jurassic rocks before being terminated in late Triassic rocks. Woodland (1971, pp. 11–35) has given a detailed log: this is summarised below.

	Thickness m	*Depth* m
Quaternary		
RECENT AND PLEISTOCENE (see Figure 37)		
Sand, silt and gravel	8.69	8.69
Boulder Clay	42.66	51.35
Varved clay and silt	12.81	64.16
Boulder Clay	7.06	71.22
Varved clay with boulder at base	6.25	77.47
Tertiary		
OLIGOCENE AND ?MIOCENE		
Lignite and Clay Unit		
Terrigenous sequence dominated by lignitic and brown clays (none over 1.5 m thick) alternating with thick units of fine silt and clay with some horizontal bedding. Some thin beds of sand and grit. Reduced iron gives general olive-green colour to fresh core	249.33	326.80
'Transition Series'		
Silt and clay, both in reduced and oxidised condition, and with sporadic lignitic partings. Below 406 m these alternate with increasingly thick units of sandstone, grit and conglomerate	114.20	441.00
Basal Red Unit		
Structureless red silt, oxidised, extensively bioturbated in places and interdigitating with thinner beds of grey-green clastics. Lacks significant organic carbon. Hematite present as small concretions near base. Thick, poorly sorted conglomerates with pebbles up to cobble size contrast with predominant fine grain size. Base sharp	160.83	601.83
Mesozoic		
LOWER JURASSIC		
Upper Lias (Toarcian Stage)		
Weathered soft red, yellow and brown clay	0.20	602.03
Grey mudstone, silty, calcareous, a few thin silty limestones and paler grey calcareous silts. Angular chert fragments in sandy limestone from 626.59 to 627.38 m	261.47	863.50
Middle Lias (Upper Pliensbachian Substage)		
Grey mudstone, silty, calcareous, thin limestone and laminated siltstone; thin conglomerate (0.05 m) at 907.03 m	147.22	1010.72
Lower Lias (Lower Pliensbachian Substage, Sinemurian and Hettangian Stages)		
Grey mudstone and siltstone, calcareous; many siltstones are intensively bioturbated. Thin ferruginous limestone and ironstone nodules between 1369 and 1468 m; below this to 1825 m, more massive limestone beds are common; to 1887 m limestone beds are thinner and less common, but thin limestone alternating with calcareous mudstone are then common down to sharp base	896.06	1906.78
UPPER TRIASSIC		
Carbonate Formation		
Grey-green dolomite, dolomitic limestone and breccia with layers of red-brown hematitic terrigenous siltstone, sandstone and replacement conglomero-breccia	16.86	1923.64
Terrigenous Formation		
Alternations of bedded pink-brown and dark brown calcitic sandstone, siltstone, arenaceous carbonate and shale; some burrows. These rest on pale red-brown calcitic calcarentic and oolitic sandstones with fan gravel and coarse sand. In lower part, an alternating sequence (as first part) recurs with local breccia and conglomerate	seen to 15.19	1938.83 base of borehole

Upper Triassic

The petrography and origin of these rocks was described in detail by Harrison (1971, pp. 37–72). The age was deduced as Upper Triassic by Warrington (1971, pp. 75–85) from an examination of 39 samples for spores. Most proved barren, but one, at 1917.8 m, yielded a substantial assemblage interpreted as late Rhaetian, and possibly close to the base of the Hettangian. Lower samples gave inconclusive evidence of a slightly older (?Norian) age. The sequence suggests deposition in a playa environment subject to flash floods. There was a reduced input of terrigenous material in the

Carbonate Formation, with signs of increasing salinity and multiphase dolomitisation as the area subsided below the Triassic sea level, allowing the formation of the marine calcareous mudstone and limestone of the *Psiloceras planorbis* Zone of the Hettangian which follows unconformably.

Lower Jurassic

An outline of the biostratigraphy based on the ammonites from this very thick sequence of argillaceous and finely arenaceous sediments was given in Ivimey-Cook (1971, pp. 87–92). The presence of 1305 m of marine sediments indicates a major Jurassic basin that has later been proved to extend far to the south-west under Cardigan Bay. The Lias sequence is about three times as thick as any other section of these beds onshore in the British Isles, and is one of the thickest sequences of this facies in Europe. The alternations of mudstone and siltstone contain virtually no coarse clastics and no clear evidence (e.g. slump structures) of proximity to a shoreline or contemporaneously active fault scarp. The rate of accumulation in different zones differs markedly; most are extremely thick, but others are thinner than in other areas (Ivimey-Cook, 1971, figs. 14–17).

Detailed studies on the distribution, taxonomy and ecology of the foraminifera of the Lower and Middle Lias by Copestake (1978), and others by Johnson (1975, 1977) on the Upper Domerian and Toarcian are summarised in Copestake and Johnson (1981) and in a detailed paper currently in press. Data from these authors are also included in Haynes (1981). The ostracods of the upper part of the sequence (*davoei* to *levesquei* zones) were studied by Sherrington (*in* Lord, 1978, p. 196).

Tertiary

The original log was prepared by O'Sullivan (1971). A discussion of aspects of the clay mineralogy and geochemistry, and conclusions on the environment were given in O'Sullivan (1979, pp. 1–13), where the sequence of fining upwards cycles are related to accumulation in meandering river channels and flood plains. These were associated with a river flowing southwards along the eastern side of the Mochras Trough over a plain of low relief, and paralleling a rather subdued fault-scarp (op. cit., p. 12). Herbert-Smith (1979, pp. 15–29) reviewed her earlier conclusions (Herbert-Smith, 1971a, 1971b) from the palynology of the sequence, and concluded that the ranges of the characteristic taxa indicate a Middle Oligocene to Early Miocene age for these deposits.

Wilkinson (1979), Wilkinson, Bazley and Boulter (1980) and Wilkinson and Boulter (1980) reviewed the Oligocene pollen and spores from the western part of the British Isles, including N. Ireland and the Bovey Basin, and placed the Mochras section as between early and late Oligocene. They suggest that the plain was only sporadically flooded, preventing the growth of a rich flood-plain flora.

Quaternary

A detailed log of this part of the sequence is given by O'Sullivan (1971) and O'Sullivan and others (1971, pp. 13–14). This is summarised in Figure 37 and put into a regional setting in Chapter 9. HCI-C

CHAPTER 9

Quaternary

During each of the several climatic deteriorations that characterised the Quaternary, local ice caps probably developed on the Welsh mountains and formed an obstruction to a major ice-sheet that advanced southwards along the Irish Sea, causing this to bifurcate on to the Cheshire Plain to the east and across the Lleyn peninsula to the west. These ice advances alternated with periods of climatic amelioration that produced temperatures at least as high as those of today, while minor fluctuations within glacial episodes caused the ice margin to advance and recede. The tills produced by the local and Irish Sea ice can be identified by their erratic suites. Local or Welsh tills are typically mid-grey and contain abundant erratics of Lower Palaeozoic rocks; the Irish Sea tills have a wider range of erratics including Carboniferous limestones, Permo-Triassic sandstones, Jurassic limestones, Cretaceous flints and Tertiary lignites, as well as Lower Palaeozoic debris. On land, however, it is difficult to discriminate between tills of the same provenance formed during different glaciations, but it has been possible locally in the Irish Sea.

IRISH SEA SEQUENCE

In the Irish Sea a firm stratigraphy for the Pleistocene deposits is now being established. Garrard (1977) has shown that two tills, both of Irish Sea lithology, are separated by inter-glacial sands and gravels which carry a rich fauna of possible Ipswichian age. The lower till, proved in BGS boreholes in outer Cardigan Bay, is the thicker and more extensive of the two, and has been tentatively correlated by Garrard with till on the Scillies; it is, however, absent from inner Cardigan Bay (Figure 35). It belongs to one of the pre-Ipswichian glaciations. The upper till averages 30 to 40 m in thickness in Cardigan Bay and St George's Channel and this thins to a southern limit some 90 km SW of Pembrokeshire in the Celtic Sea (BGS unpublished current work). It, too, is absent from the eastern part of Cardigan Bay, for it meets the Welsh till on the seaward side of the drift ridges known as Sarn Badrig, Sarn-y-Bwch and Cyn felin Patches (Figure 36). It is almost certainly late-Devensian in age, and is variously dated at 20 000 BP (Mitchell, 1972) or 18 000 to 14 468 BP (Rowlands, 1980). Bowen (1977) believed that Sarn Badrig is the terminal moraine of the 'Upper Boulder Clay' (Llanystumdwy Till). Garrard maintained, however, that the Sarns are composed of Welsh till from piedmont glaciers, and mark the interfluve between major onshore valleys.

A local re-advance of the Devensian Irish Sea ice, the 'Second Recorded Advance' of Saunders (1968), reached a position south of Bardsey Island and deposited a terminal moraine along the north-west of the Lleyn peninsula from Pen-y-groes to Trevor. In Tremadoc Bay, Welsh drift overlying Irish Sea drift may mark the corresponding latest advance of the Welsh ice.

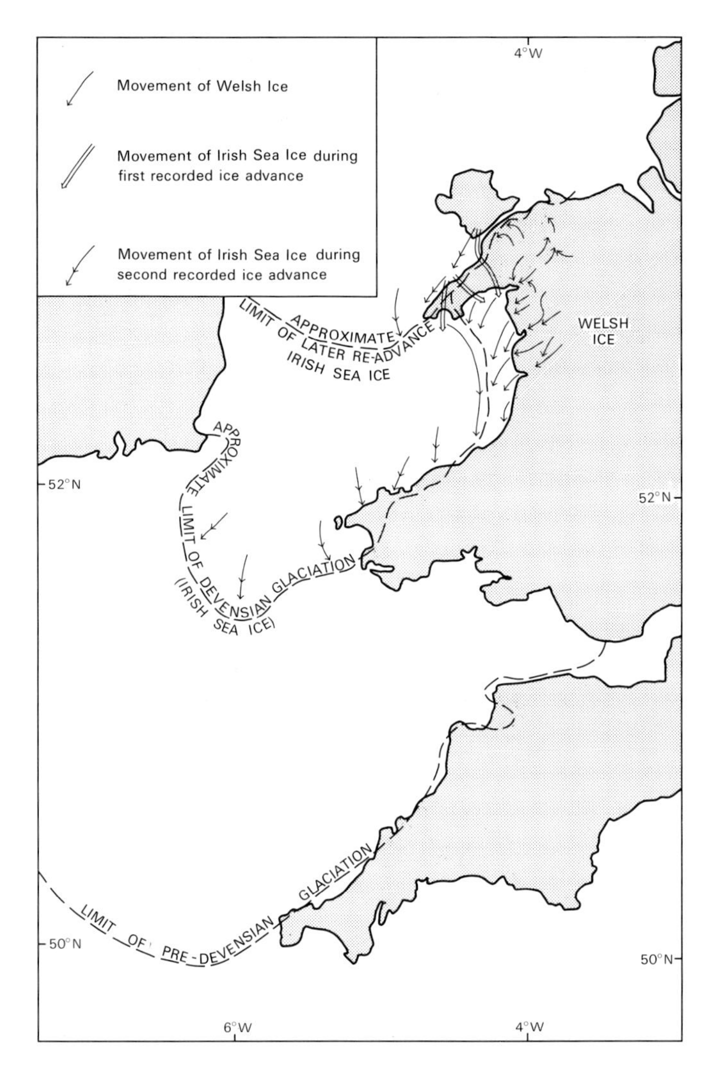

Figure 35 Limits of Pleistocene glaciations compiled from Garrard (1977) and Saunders (1968)

The latest Devensian and Flandrian history of the area is not yet fully understood. Recent offshore exploration has shown the presence of deep-channels in Cardigan Bay. Pinger profiles show that they have had a complex history with several periods of sedimentation separated by episodes of further channelling. Muddy Hollow, central to Tremadoc Bay north-west of Sarn Badrig, continues the line of the Glaslyn/Dwyryd estuary, and this deep channel which is buried under Morfa Harlech may have continued in use throughout much of the latest Quaternary. In the Mawddach estuary superficial deposits have been proved to 23 m in boreholes, and geophysical evidence suggests that there are three adjacent channels varying in depth between 24 and 43 m. In order to cut such deep channels subaerially, Blundell, King and Wilson (1964) estimated that the sea

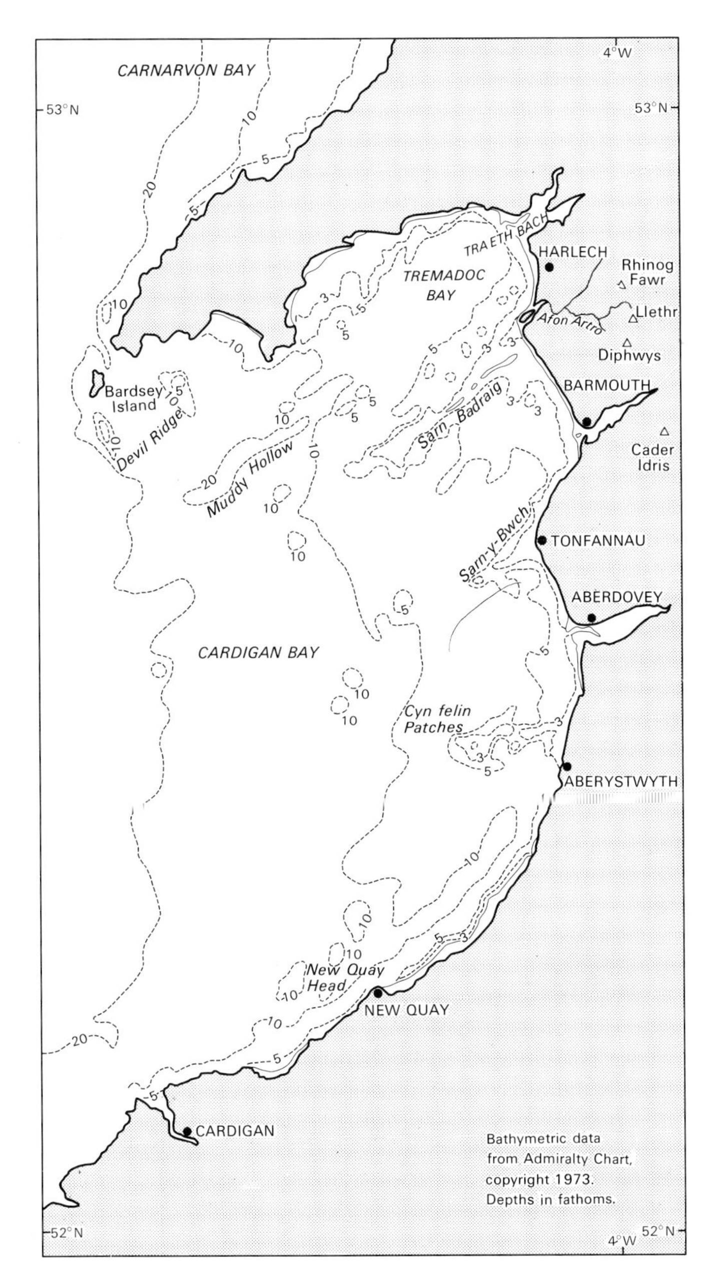

Figure 36 Sketch map showing sea-floor topography in Cardigan Bay

level must have stood some 90 m below present OD. These channels are filled by laminated argillaceous sediments. Towards the shore peat lies above the buried channels, with a fauna diagnostic of estuarine and lagoonal conditions. The associated sediments become more marine in a seaward direction and also upwards, recording the progress of the Flandrian marine transgression which reworked the surface of the underlying till to produce a thin lag deposit. A radiocarbon date (Haynes and others, 1977) from peat collected near the top of a buried channel in Trawling Ground, off Newquay Headland (18.5 m below present OD), has given an age of 8740 ± 100 BP, (Late Glacial or Devensian Zone III): clearly most of the channel fill preceded this date. The isostatic rise in sea level following de-glaciation reached a peak by 5000 BP (Atlantic Stage, Zone VII), while some isostatic adjustments still continues (Synge, 1977). Along the shores of the Irish Sea, this adjustment is expressed as raised beaches to the north, and as a drowned coastline to the south. Traeth Bach and the Mawddach estuaries on the Harlech coast represent drowned river valleys, and the present day coastline is predominantly one of accretion.

MOCHRAS FARM BOREHOLE

The Mochras Farm Borehole (Woodland (Editor), 1971), is the only on-land boring that establishes the presence of more than one till (Figure 37). The authors report as follows:

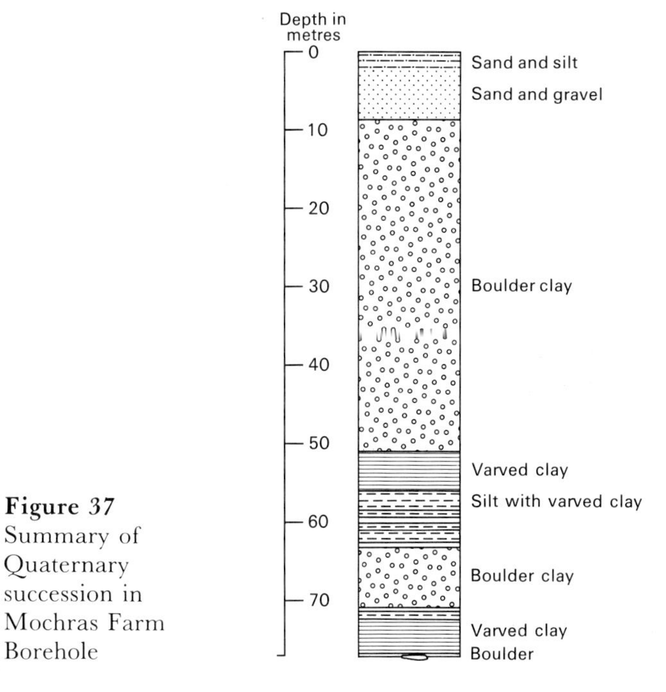

Figure 37 Summary of Quaternary succession in Mochras Farm Borehole

'A 60-cm fully cored boulder of local grit abruptly marks the base of the Pleistocene sequence at 77.47 m. The boulder is immediately overlain by 2.5 m of variously slumped and horizontally varved clays, which in turn is overlain by 10 m of what appears to be rather fine boulder clay, interrupted in the centre by a thin sequence of broken varves and horizontal fine silt laminae. This alternation is continued by a sequence of finely banded silts with moss partings, capped by contorted varved clays, the unit being 8 m thick. A boulder clay 4.8 m thick is followed by 1 m of contorted varved clays before a final sequence of 46 m of boulder clay, most of which was not cored, but which may be stratified in some fashion as the drillers report some more easily drilled layers.'

'Some of the contortions of the varved clays are no doubt due to ice-drag, but in other cases some very clear slump structures are recognised, very similar to those described by A. J. Smith (1959) from Lake Windermere.'

'At the top of the borehole 6.5 m of Post-Pleistocene sediments were also not cored; they appear to consist almost entirely of sand, silt and shingle with shells of recent affinity, and the nature of the contact with the Pleistocene beds below is unknown.'

Despite the coastal position of the borehole, and the nature of the sequence, which in many ways resembles that found farther off-shore, the clasts so far recorded from the tills are all of local origin.

Palynological analysis of the cores from 46.9 to 77.47 m (i.e. below the main part of the 'upper' till) was carried out by Herbert-Smith (1971b). Pollen, spores, a diverse moss fauna, scattered leaves and 'hystrichospheres' were recorded. They do not, however, allow of precise dating, and merely point to an interstadial or interglacial episode separating the two main till sequences.

LAND SEQUENCE

During the late Devensian the Harlech district was covered by local Welsh ice forming piedmont and valley glaciers. Radiocarbon datings by Lowe (1981) on lake sediments that formed after this ice melted in Llyn Gwernau, on the north side of Cader Idris, are dated at 13 200 ± 120 years BP, thus estabishing the minimum age for the late-Devensian melt. No evidence has yet been acquired on land of any earlier glaciation. An ice-shed, corresponding approximately with the present-day water-shed, lay to the west of the Arenigs and east of Rhobell Fawr (Foster, 1968; Rowlands, 1970); ice moved eastwards to the valley of the Dee and westwards towards the Irish Sea. From a detailed study of the erratics and boulder-clay fabrics of this Merionethshire ice-sheet, Foster concluded that the Rhinogs deflected the bulk of the lower ice layers to the north-west and the south, while some of the ice passed through transfluence cols and breached the escarpment. The upper ice layers flowed completely over the Rhinogs into Cardigan Bay. Thus, distinctive erratics from the Rhobell Fawr area and from the east of the Harlech Dome were deposited to the west of the Rhinog escarpment, an area which is otherwise occupied by more local tills. The higher ground of the Rhinogs was subjected to intense scouring by relatively fast moving ice. In the northern part of the Rhinogs ice-scoured pavements have been preserved on the Rhinog Formation and are littered with large erratics, but softer weathering shale formations predominate southwards from Y Llethr and the slopes here are mantled with the products of mass wasting.

One of the most conspicuous features of the present landscape is a series of intersecting, steep-sided, dry valleys. The sides of these deep hollows are commonly marked by accumulations of scree of head, and some of the floors contain a drift-fill punctuated by small round depressions filled with peat. The dominant trends of the hollows are approximately NNW, N and ENE, corresponding roughly with the major joint directions (Figure 38). The major hollows cross the present water-sheds without apparent deviation. The smaller hollows which criss-cross the area may follow major joints which have been preferentially scoured by sub-glacial drainage channels. Those hollows which follow the main flow direction of the ice are overdeepened, in comparison with their N- and NNW-trending counterparts.

Other dry valleys occur in all parts of the district; they are probably abandoned meltwater channels that were active when the main drainage was blocked by ice. One such channel, floored by alluvium, runs about 1.5 km along Bwlch-y-goedleoedd, west of Panorama Walk at Barmouth. The NNE-trending valley north of Nannau Hall [743 208] may have a similar origin.

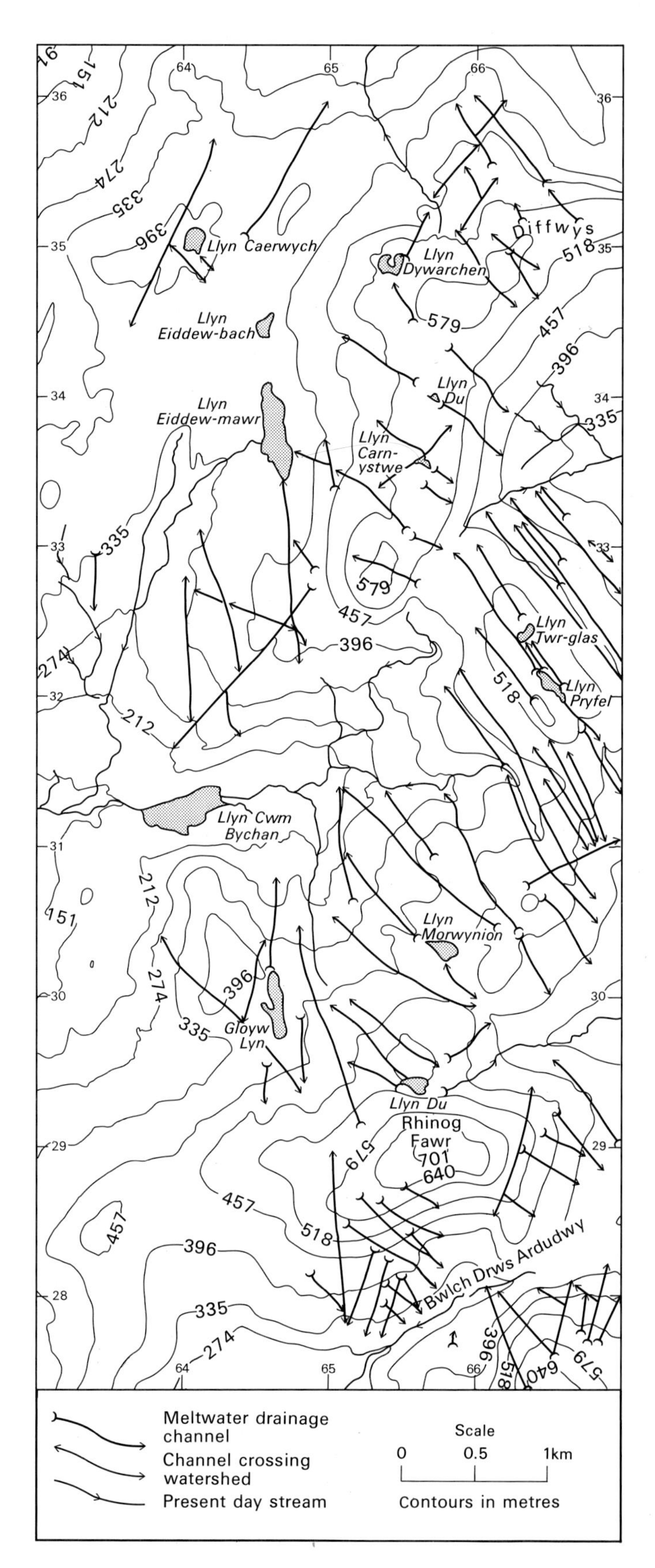

Figure 38 Sketch map of part of the Rhinog mountains showing the distribution and orientation of possible sub-glacial drainage channels following main joint directions

QUATERNARY DEPOSITS

Moraine

Though the surface topography of most of the glacial landforms appears to have been subdued or been flattened by solifluction, some moraines can still be identified on the valley sides (for example in Cwm Bychan). South of Llanbedr, a lateral moraine marks the junction between the valley glaciers that occupied the Dwyryd and Ysgethin valleys. Its eastern end has three distinct linear crests, but the feature is broken into more rounded hills to the west. This moraine ridge (Plate 13.3) is in line with Sarn Badrig, the sea-rise in Cardigan Bay which is exposed at low tide (Figure 36). From bathymetric charts it can be seen that Sarn Badrig is also multi-crested.

On the south side of Afon Ysgethin a rather subdued moraine has a single crest-line composed largely of boulders. The ridge stretches east from the cairns [6116 2201] into the valley. A NE-trending continuation of this feature forms the barrier for Llyn Irddyn. South of the moraine, and running parallel with it, a dry valley may represent an abandoned marginal meltwater channel which drained the side of the Ysgethin glacier.

Boulder Clay

After the melting of the ice cap, boulder clay, deposited mainly as lodgment till, covered the lower slopes of the Rhinogs and the valleys. The main valley-fill reaches a height of about 400 m, with only comparatively small tongues preserved in the upland cols. There are few natural sections through the boulder clay, but locally on river banks up to 6 m of till are exposed, and 30 m have been recorded in boreholes around Capel Hermon (Rice and Sharp, 1976).

The boulder clay is typically grey to olive in colour, weathering to brown, but locally takes on the colour of the underlying beds. This is particularly noticeable on the Dolgellau Member where the clay is almost black. Erratics are entirely of local origin. The only exception is in the boulder clay which flanks the Dwyryd, south-east of Talsarnau, where rare fragments of Carboniferous limestone have been recorded. South-westwards, between Harlech, Llandanwg and Moelfre [626 246], the till is a distinctive reddish brown, and unsorted sandy gravel replaces the more normal clayey matrix. The erratics are predominantly local with some exotics, and this till may have been deposited by the Dwyryd/Glaslyn glacier.

Blocky tills containing little or no clayey matrix occur on the higher slopes and in the valley heads. This material was carried on or within the ice, and dumped when the ice melted. A good example of the hummocky terrain so produced occurs at the head of a valley [669 356] between Diffwys and Moel Y Gyrafolen. Here, chaotic mounds of bouldery till hold small ponds of water in the intervening depressions.

Drumlins are uncommon even on the wider tracts of boulder clay. A few isolated ones occur on the western side of the Rhinogs, on the slopes south of Talsarnau and east of Tal-y-bont. To the east of the Rhinogs, drumlin swarms occur in the valleys of the Prysor, the Wnion and the Mawddach near its junction with Afon Ceirw. In the Prysor a number of drumlin-like features dissected by the cuttings for the old Bala/Ffestiniog railway have solid cores. A group of drumlins south of Trawsfynydd lake are semi-circular in plan and may be the reworked deposits of an earlier mountain valley end-moraine.

AAJ

Fluvioglacial Terrace Deposits

A valley train deposit occurs in the Mawddach valley near Ganllwyd, between the Ty'n-y-groes hotel [729 232] and the confluence of the Eden and Mawddach. The deposit, about 12 m thick, consists of locally-derived rounded boulders and cobbles in a matrix of sand and gravel, with ochre and clay in places and lenses of coarse lithic sand. Parts of the deposit

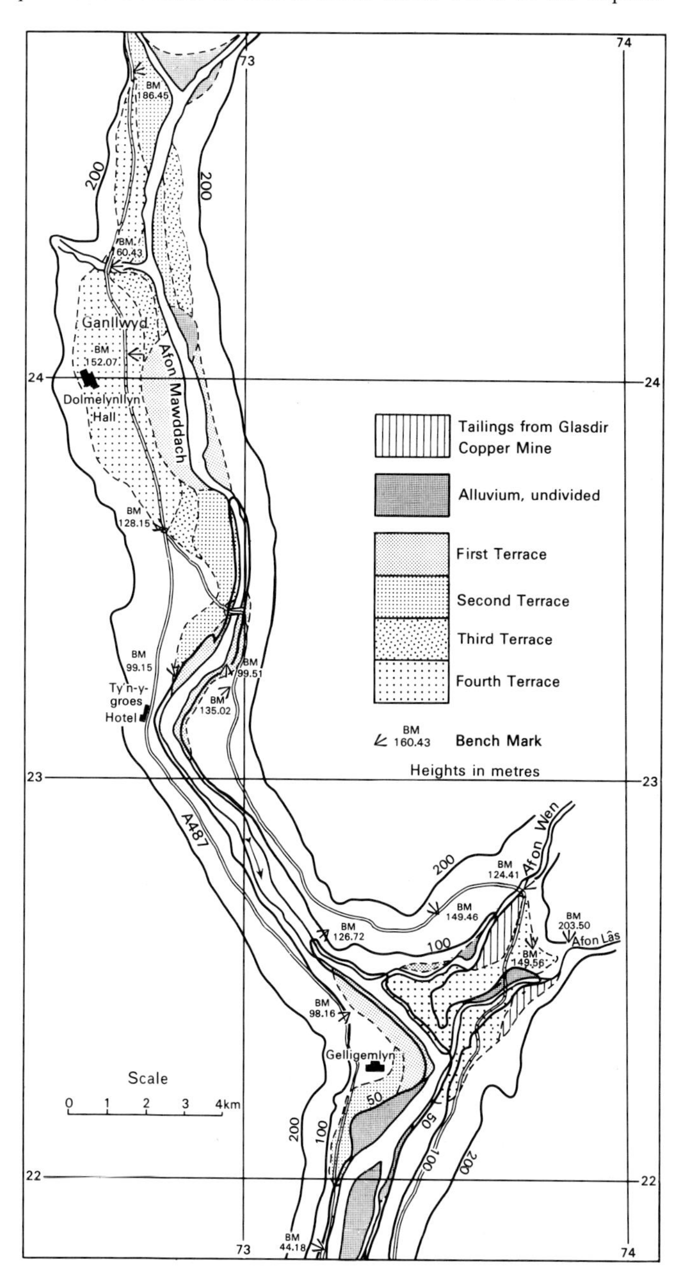

Figure 39 Terraces in fluvioglacial deposits in the Mawddach valley

have a hard manganiferous or, more commonly, ferruginous cement. A similar deposit forms a hog-back dissected by the Wen and Lâs [735 224] where they join the Mawddach about 1.5 km below the Ty'n-y-groes; it is present on both sides of the Mawddach below this. The terracing of this deposit is unique in the area (Figure 39), and is probably the result of aggradation in the Mawddach with the rapid deposition of outwash material. PMA

Fluvioglacial Gravel

At a locality in the railway cutting north of Llandanwg, a lens of gravel is overlain by about 6 m of boulder clay, which in turn is overlain by up to 4.5 m of cross-bedded sandy gravel. The upper gravel forms a low feature which can be traced uphill into a dry channel and is probably a proglacial deposit laid down by meltwater. To the north-east this channel passes into a marked shoulder in the boulder clay. There is a second dry channel, subparallel to this, to the south-east, and both probably carried meltwaters from the edge of the Dwyryd/Glaslyn glacier. Elsewhere, a few isolated mounds of gravel rest on the boulder clay. They have generally been worked, despite the small volume of material. One occurs on the left bank of Afon Clywedog [7604 1783]; another lies south of Trawsfynydd lake [7040 3344], to the south-east of an elongated drumlin, and shows arched bedding. The form of this mound suggests that it may have originated as a kame. AAJ

Sections through boulder clay in places reveal thin lenses of sand and gravel of subglacial origin. Such an undermelt drift sequence is exposed in a cutting by the road along Nant Cesailgwm [6907 2007] where about 5 m of sandy cobbly gravel are overlain by 1 m of brown boulder clay with lenses of lithic sand near the bottom. Sequences like this were probably laid down beneath slowly melting, stagnant ice.

In a road cutting near Llanelltyd [7218 2050], a section across a stream channel (Plate 14.3) that is cut into bedrock on the steep valley side is filled with a chaotic mixture of angular boulders, sand, and gravel with a ferruginous cement. This deposit was formerly covered by scree, which was stripped away when the road was widened in 1976–79. Another similar deposit fills a dry valley and is partly covered by scree at the roadside [6797 1890] 1 km E of Bontddu. Both are likely to be old stream courses, reactivated and filled with outwash during deglaciation. PMA

Scree and Head

Deposits of scree mantle most of the steep slopes and, though they were initiated in periglacial conditions, some are still active, e.g. at Cwm Yr Allt-lŵyd and below Precipice Walk along the Mawddach valley. Deposits of head are most conspicuous where they rest on mudstone and siltstone formations, and are extensive on the Rhinogs south of Y Llethr, where the highest parts of the ridge are on the Hafotty and Gamlan formations.

The slopes to the west of Diffwys [6612 2343] and south-west of Rhinog Fawr may be altiplanation terraces. These features, cut in bedrock, lack a dominant structural control, and are the result of localised erosion due to frost action and mass wasting. Gelifluction deposits mantle the higher slopes, and north-west of Llyn Bodlyn a lobe of this material completely overrides the escarpment of the grits of the Barmouth Formation. On the slopes of Ffridd Hari Howel [760 320], a series of terraces in the boulder clay may represent the edge of old gelifluctcd benches. Small block fields, the result of *in situ* weathering (Washburn, 1973), commonly occur near the junction of solid and drift. The effect of hill creep on the underlying solid rock can be seen in a quarry (Plate 14.1) on the flanks of the Moel Hafod Owen [7583 2661] where the direction of dip has been reversed.

Other remnants of a periglacial regime are apparent. Polygonal patterned ground is discernible in places, and stone stripes can be seen on the slopes of Rhinog Fach (Goodier and Ball, 1969). Frost heaved blocks (Washburn, 1973) are common on the ice-scoured pavements on the northern part of the Rhinog escarpment, e.g. Diffwys [665 352].

Storm Gravel Beach, Marine and Estuarine Alluvium

The coastline from Harlech to Barmouth is essentially one of accretion, with the build-up of estuarine sediments at the mouth of the Glaslyn extending the area of Morfa Harlech. Sediment swept from the Mawddach contributes to that from the Ysgethin and Artro rivers to make up the Morfa Dyffryn complex. Active coastal erosion appears to be confined mainly to the boulder clay cliffs at Llandanwg and Mochras Island.

The Dwyryd is stablised in its present position by artificial embankments, but there has been much change in its course even in historic times. A small patch of storm-beach deposits [6022 3300] marks the position of a former shoreline. Harlech Castle, built on the coast in the 12th century and now almost a kilometre from the sea, gives some indication of the age of this deposit. A more recent channel of the Dwyryd to the south-east of Ynys Llanfihangel-y-traethau is now marked by peat-filled depressions and ox-bows. Lithologically Morfa Harlech seems less variable than Dyffryn, and consists predominantly of grey clay with some sandy clay: a thick layer of blown sand covers the western part of the Morfa. The sand in the Traeth is beige and well washed, but bands of black euxinic sand are present at the mouth of the channel and in the back swamps. Brownish grey mud, which thickens inshore, collects on the tidal flats (Plate 14.2), and this is probably the derivation of most of the clay seen on the Morfa.

Short stretches of storm beach gravels occur south of Llandanwg and east of Mochras. A more extensive deposit, several kilometres long, stretches along the seaward edge of the dunes on the southern part of Morfa Dyffryn. The gravel consists of rounded boulders and cobbles of both local and exotic lithologies.

At the northern end of Morfa Dyffryn, Mochras Island is a boulder clay ridge now connected to the mainland by the build-up of estuarine sands. The sequence seen in shallow hand-auger holes and drains on the Morfa shows layers of laminated clays, sand and intercalated peats.

Alluvial Fan

Small areas of alluvial fan occur where mountain streams

Plate 13

1 View north over the dunes on Morfa Harlech with the Moelwyn Hills in the distance. (L 1337).

2 Landslip in boulder clay at Cwmhesgen (L 1330)

3 Morainic ridge, west of the Rhinogs, running seawards towards Sarn Badrig (L 1331).

Plate 14

1 Hillcreep in Ffestiniog Flags Formation in an old quarry on Moel Hafod Owen. (L 1335).

2 Erosion of silty mud layer above sand near Portmerion, north side of Traeth Bach. (L 1336).

3 Partly cemented bouldery gravel in an old, possibly subglacial stream course recently revealed by the removal of scree on west side of Mawddach valley, near Llanelltyd. (L 1333).

undergo an abrupt change in gradient, for example where the streams join the main valley or where the stream course passes from steep solid ground to more gentle drift covered slopes as on the north side of Moelyblithcwm [650 259]. Two fans [7955 2900 and 797 288] occur in Cwm yr Allt-lŵyd where the streams flow off the Ordovician escarpment on to the Cambrian. The sediments are locally derived pebbles and cobbles in a silty sand matrix. AAJ

River Terrace Deposits and Alluvium

The three major rivers, the Mawddach, Wnion and Prysor, flow in deeply glaciated valleys, now floored by alluvium, with scree-covered sides. The present-day watershed corresponds essentially to that operating during the Late Glacial as described by Foster (1968) and Rowlands (1970). With minor modifications the main drainage system is essentially pre-glacial. Modifications that have been taken place are largely because of capture resulting from the blockage of valleys by drift. Afon Prysor probably originally flowed into the Eden and was captured by Ceunant Llennyrch because of a boulder clay blockage south of Llyn Trawsfynydd. According to Rice and Sharp (1976), the upper part of Afon Wen once flowed into the Mawddach about 4 km above its present confluence. The original river valley was blocked around Capel Hermon [746 256] by boulder clay, causing the river to turn southwards into its present course.

Narrow strips of alluvium are present along most of the rivers and streams. The most extensive is at the confluence of the Wnion and Mawddach at Llanelltyd, where borings have shown that clay, silt and peat in most places overlie gravel. In the lower part of the Wnion valley, the gravel at one locality [7383 1795] is known to be 27 m thick. In narrow upland valleys the alluvium is commonly coarse lithic sand and gravel; it is not uncommon to find layers of hard manganiferous and ferruginous pan within it; e.g. in Afon Gain [7529 3295] and Afon Lliw [8008 3392]. In addition to the alluvial flood plain, two terraces can be recognised in the valleys of some of the larger streams; for example on the Wnion south of Rhydymain and at the confluence of Nant Hîr and Afon Gain [748 324]. PMA, AAJ

Lacustrine Deposits

Rock-basin lakes, caused by differential glacial erosion in valleys or along joint lineaments, are numerous; they include Llyn Cwm Bychan, Llyn Morwynion, Llyn Eiddew-mawr and Gloyw Lyn. Many others are now filled with lacustrine gravels and peat. A number of these old lake basins occur in the catchment area of the Artro. In Cwm Nantcol two large basins blocked by rock bars [6180 2676, 6342 2618] are filled with peat, and peat on lacustrine alluvium respectively.

On the north slope of Y Llethr, the lakes Perfeddau and Hywel and Llyn Y Bi are classic cirque-lakes, separated by arêtes and blocked by terminal moraine debris. Farther north, another cirque is occupied by Llyn Dywarchen [6544 3490], but in some of the other valleys lakes have not survived. At the head of the Ysgethin the highest cirque is filled with peat [6654 2434], and on the north slopes of Moel Y Gyrafolen the head of the cirque is occupied by flow-till.

Till-dammed lakes are less common. In the Ysgethin valley Llyn Irddyn is held between the valley side and a lateral moraine, and Llyn Bodlyn and Llyn Dulyn behind a recessional cross-valley moraine. South-west of Moel Geodog [6134 3252] some relatively large basins between drumlins are filled with gravel underlying peat [eg. 605 314]. Llyn Cynwch [737 206], situated in a valley that may originally have been a subglacial channel, is blocked at both ends by till. A few small round basins, mainly filled with peat, may be the remnants of pingos or kettle-holes [eg. 5920 3010].

Peat

Peat occurs in rock basins and commonly marks the sites of former lakes but, in addition, large areas of boulder clay and solid are covered by peat in the upland regions. The peat contains fragments of birch, and thicknesses up to 2.5 m are common.

On the low ground of Morfa Harlech and Morfa Dyffryn peat occurs in abandoned ox-bows and intercalated with the marine and estuarine deposits which are now partially concealed by blown sand. The peat contains grasses, sedges and abundant full-sized tree trunks, and is currently being eroded on the beach south of Tal-y-bont. An almost continuous pod of thick peat occurs at the back of Morfa Dyffryn in front of the boulder clay bank, and on Morfa Harlech thick peat occupies a similar position at the back of the estuarine flats but the deposit is less continuous on the surface.

Blown Sand

A ridge of well defined sand dunes skirts the coast along the western ridge of Morfa Harlech and Morfa Dyffryn. On Morfa Harlech the dunes (Plate 13.1) rise locally to 15 m OD, tapering away in the north towards the estuary. Thin patches of peat occurs in the hollows between the dunes. Inland blown sand covers the western part of the estuarine alluvium as well as the higher ground around Ynys Llanfihangel-y-traethau and Ynys Gifftan.

On Morfa Dyffryn the dunes have been flattened for the runway at Llanbedr airfield, and to the south, most of the estuarine clays are covered by blown sand.

Landslips

There are many small landslips in boulder clay along the valley sides, formed where the slopes have been oversteepened by stream erosion, for example in Cwm Hesgen [7865 2945] (Plate 13.2). Site investigation boreholes along the eastern part of the Dolgellau by-pass proved 7 m of slipped boulder clay overlying Recent alluvial sediments along the northern edge of the Wnion. AAJ

REFERENCES

ALDER, K. E. 1976. The geology of the area around Bettws Gwerfil Goch and Melin-y-wig. Clwyd. Unpublished PhD thesis, University of Cambridge.

ALLEN, P. M., COOPER, D. C., FUGE, R. and REA, W. J. 1976. Geochemistry and relationships to mineralization of some igneous rocks from the Harlech Dome, North Wales. *Trans. Instn. Min. Metall.*, Sect. B, Vol. 85, 100–108.

— — and SMITH, I. F. 1979. Mineral exploration of the Harlech Dome, North Wales. *Mineral Reconnaissance Programme Rep. Inst. Geol. Sci.*, No. 29.

— and EASTERBROOK, G. D. 1978. A mineralised breccia pipe and other intrusion-breccias in the Harlech Dome, North Wales. *Trans. Instn. Min. Metall.*, Sect. B, Vol. 87, 157–161.

— and JACKSON, A. A. 1978. Bryn-teg Borehole, North Wales. *Bull. Geol. Surv. G.B.*, No. 61, 51 pp.

— — and RUSHTON, A. W. A. 1981. The stratigraphy of the Mawddach Group in the Cambrian succession of North Wales. *Proc. Yorkshire Geol. Soc.*, Vol. 43, 295–329.

ANDREW, A. R. 1910. The geology of the Dolgelley Gold Belt, North Wales. *Geol. Mag.*, Dec. V., Vol. 7, 159–171, 201–211, 261–271.

ARCHER, A. A. 1959. The distribution of non-ferrous ore in the Lower Palaeozoic rocks of North Wales. Pp. 259–276 in *The future of non-ferrous mining in Great Britain and Ireland.* Symposium. *Instn. Min. Metall.*

BASSETT, D. A. 1954. The folds and cleavage of the Talerddig district of West Montgomeryshire. *Adv. Sci.*, Vol. 11, No. 41, 108–110.

— WHITTINGTON, H. B. and WILLIAMS, A. 1966. The stratigraphy of the Bala district, Merionethshire. *Q.J. Geol. Soc. London*, Vol. 122, 219–271.

BASSETT, M. G., OWENS, R. M. and RUSHTON, A. W. A. 1976. Lower Cambrian fossils from the Hell's Mouth Grits, St Tudwal's Peninsula, North Wales. *J. Geol. Soc. London*, Vol. 132, 623–644.

BATES, D. E. B. 1965. A new Ordovician crinoid from Dolgellau, North Wales. *Palaeontology*, Vol. 8, 355–357.

— 1975. Slaty cleavage associated with sandstone dykes in the Harlech Dome, North Wales. *Geol. J.*, Vol. 10, 167–175.

BECKINSALE, R. D. and RUNDLE, C. C. 1980. K-Ar ages for amphibole separates from the Rhobell Volcanic Group (Upper Tremadocian) Harlech Dome, North Wales. *Rep. Inst. Geol. Sci.*, No. 80/1, 9–11.

BELT, T. 1867. On the 'Lingula Flags' or 'Festiniog Group' of the Dolgelly district. Parts I and II. *Geol. Mag.*, Vol. 4, 493–495, 536–543.

— 1868. On the 'Lingula Flags' or 'Festiniog Group' of the Dolgelly district. Part III. *Geol. Mag.*, Vol. 5, 5–11.

BLUNDELL, D. J., DAVEY, F. J. and GRAVES, L. J. 1968. Sedimentary basin in the South Irish Sea. *Nature, London*, Vol. 219, 55–56.

— — — 1971. Geophysical surveys on the South Irish Sea and Nymphe Bank. *J. Geol. Soc. London*, Vol. 127, 339–375.

— GRIFFITHS, D. H. and KING, R. F. 1969. Geophysical investigations of buried river valleys around Cardigan Bay. *Geol. J.*, Vol. 6, 161–181.

— KING, R. F. and WILSON, C. D. V. 1964. Seismic investigations of the rocks beneath the northern part of Cardigan Bay, Wales. *Q.J. Geol. Soc. London*, Vol. 120, 35–50.

BOWEN, D. Q. 1977. The coast of Wales. Pp. 223–256 in *The Quaternary History of the Irish Sea.* KIDSON, C. and TOOLEY, M. J. (Editors). (Liverpool: Seel House Press.)

BROMLEY, A. V. 1971. Phases of deformation in North Wales. (Letter to the Editor). *Geol. Mag.*, Vol. 108, 548–550.

BULLERWELL, W. and MCQUILLIN, R. 1969. Preliminary report on a seismic reflection survey in the southern Irish Sea. *Rep. Inst. Geol. Sci.*, No. 69/2, 7 pp.

CLARKSON, E. N. K. 1967. Environmental significance of eye-reduction in trilobites and Recent arthropods. *Mar. Geol.*, Vol. 5, 367–375.

COCKS, L. R. M. 1978. A review of British Lower Palaeozoic brachiopods, including a synoptic revision of Davidson's Monograph. *Monogr. Palaeontogr. Soc. London*, 256 pp. (Pub. 549, part of Vol. 131, for 1977.)

COLE, G. A. J. and HOLLAND, T. 1890. Structure and stratigraphical relations of Rhobell Fawr. *Geol. Mag.*, Dec. 3, Vol. 7, 447–452.

COOPER, D. C., CAMERON, D. G., JACKSON, A. A., ALLEN, P. M. and PARKER, M. E. 1983. Exploration for volcanogenic sulphide mineralisation at Benglog, North Wales. *Mineral Reconnaissance Programme Rep. Inst. Geol. Sci.*, No. 63.

COPESTAKE, P. 1978. Foraminifera from the Lower and Middle Lias of the Mochras Borehole. Unpublished PhD thesis, University of Wales, Aberystwyth.

— and JOHNSON, B. 1981. Jurassic. Part 1. The Hettangian to Toarcian. Pp. 81–105 in *Stratigraphical atlas of fossil foraminifera.* JENKINS, D. G. and MURRAY, J. W. (Editors). (British Micropalaeontological Society: Ellis Harwood Ltd.).

— — and STUBBLEFIELD, C. J. 1972. A correlation of Cambrian rocks in the British Isles. *Spec. Rep. Geol. Soc. London*, No. 2, 42 pp.

COX, A. H. 1925. The geology of the Cader Idris range (Merioneth). *Q.J. Geol. Soc. London*, Vol. 81, 539–594.

— and LEWIS, H. P. 1945. Summer field meeting, 1944. Dolgelly district. *Proc. Geol. Assoc.*, Vol. 56, 59–81.

— and WELLS, A. K. 1915. The Ordovician sequence in the Cader Idris district (Merioneth). *Rep. Br. Assoc. Sci. for 1915 (Manchester)*, 424–425.

— — 1921. The Lower Palaeozoic rocks of the Arthog–Dolgelley district (Merionethshire). *Q.J. Geol. Soc. London*, Vol. 76, 254–324.

— — 1927. The geology of the Dolgelley district, Merionethshire. *Proc. Geol. Assoc.*, Vol. 38, 265–318.

CRIMES, T. P. 1970. A facies analysis of the Cambrian of Wales. *Palaeogeography, Palaeoclimatol., Palaeoecol.*, Vol. 7, 113–170.

DAVIES, R. G. 1959. The Cader Idris granophyre and its associated rocks. *Q.J. Geol. Soc. London*, Vol. 115, 189–216.

DAVIES, W. and CAVE, R. 1976. Folding and cleavage determined during sedimentation. *Sediment. Geol.*, Vol. 15, 89–133.

DE SITTER, L. U. 1964. *Structural Geology.* Second edition. (New York, San Francisco, Toronto, London: McGraw-Hill Book Company.) 551 pp.

DEWEY, H. and BROMHEAD, C. E. N. 1915. Tungsten and Manganese Ores. *Spec. Rep. Miner. Resour. Mem. Geol. Surv. G.B.*, Vol. 1.

— and DINES, H. G. 1923. Tungsten and Manganese Ores. 3rd edition. *Spec. Rep. Miner. Resour. Mem. Geol. Surv. G.B.*, Vol. 1.

— and EASTWOOD, T. 1925. Copper ores of the Midlands, Wales and the Isle of Man. *Spec. Rep. Miner. Resour. Mem. Geol. Surv. G.B.*, Vol. 30, 39–48.

— and Smith, B. 1922. Lead and zinc ore in the pre-Carboniferous rocks of West Shropshire and North Wales. Part 2 - North Wales. *Spec. Rep. Miner. Resour. Mem. Geol. Surv. G.B.*, Vol. 23.

Dewey, J. F. 1969. Evolution of the Appalachian/Caledonian orogen. *Nature, London*, Vol. 222, 134-139.

Dobson, M. R., Evans, W. E. and Whittington, R. 1973. The geology of the South Irish Sea. *Rep. Inst. Geol. Sci.*, No. 73/11, 35 pp.

Dunkley, P. N. 1978. The geology of the south western part of the Aran Range, Merioneth, with particular reference to the igneous history. Unpublished PhD thesis, University of Wales, Aberystwyth.

— 1979. Ordovician volcanicity of the SE Harlech dome. Pp. 597-601 in *The Caledonides of the British Isles — reviewed.* Harris, A. L., Holland, C. H. and Leake, B. E. (Editors). *Spec. Pub. Geol. Soc. London*, No. 8.

Dzulynski, S. and Walton, E. K. 1965. Sedimentary features of flysch and greywacke. *Developments in Sedimentology*, Vol. 7. (Amsterdam: Elsevier Press.)

Evans, G. 1965. Intertidal flat sediments and their environments of deposition in the Wash. *Q. J. Geol. Soc. London*, Vol. 121, 209-245.

Fearnsides, W. G. 1905. On the geology of Arenig Fawr and Moel Llyfnant. *Q. J. Geol. Soc. London*, Vol. 61, 608-640.

— 1910. The Tremadoc slates and associated rocks of south-east Caernarvonshire. *Q. J. Geol. Soc. London*, Vol. 66, 142-188.

— and Davies, W. 1944. The geology of Deudraeth. The country between Traeth Mawr and Traeth Bâch, Merioneth. *Q. J. Geol. Soc. London*, Vol. 99, 247-276.

Fitches, W. R. 1972. Polyphase deformation structures in the Welsh Caledonides near Aberystwyth. *Geol. Mag.*, Vol. 109, 149-155.

Folk, R. L. 1974. *Petrology of sedimentary rocks.* (Austin: Hemphill Publishing Co.)

Forbes, D. 1867. Researches in British mineralogy. *Philos. Mag.*, Ser. IV, Vol. 34, 329-354.

Forster, A. 1976. Density, porosity, magnetic susceptibility, resistivity and formation factor determinations in nine samples from North Wales. *Rep. Engrg. Geol. Unit, Inst. Geol. Sci.*, No. 83. (Unpublished.)

Fortey, R. A. and Owens, R. M. 1978. Early Ordovician (Arenig) stratigraphy and faunas of the Carmarthen district, south-west Wales. *Bull. Brit. Mus. (Nat. Hist.) Geol.*, Vol. 30, 225-294.

Foster, H. D. 1968. The glaciation of the Harlech Dome. Unpublished Ph.D thesis, University of London.

Garcia, M. O. 1978. Criteria for the identification of ancient volcanic arcs. *Earth Sci. Rev.*, Vol. 14, 147-165.

Garrard, R. A. 1977. The sediments of the south Irish Sea and Nymphe Bank area of the Celtic Sea. Pp. 69-72 in *The Quaternary History of the Irish Sea.* Kidson, C. and Tooley, M. J. (Editors). (Liverpool: Seel House Press.)

Geological Survey of Great Britain. 1965. *Aeromagnetic map of Great Britain*, 1:625 000, Sheet 2. (Southampton: Ordnance Survey.)

George, T. N. 1963. Palaeozoic growth of the British Caledonides. Pp. 1-34 in *The British Caledonides.* Johnston, M. R. W. and Stewart, F. H. (Editors). (Edinburgh and London: Oliver & Boyd.)

Gilbey, J. W. G. 1969. The mineralogy, paragenesis and structure of the ores of the Dolgellau Gold Belt, Merionethshire, and associated wall rock alteration. Unpublished Ph.D. thesis, University of London.

Glasby, G. P. 1974. A geochemical study of the manganese ore deposits of the Harlech Dome, North Wales. *J. Earth Sci.*, Vol. 8, 445-450.

Goodier, R. and Ball, D. F. 1969. Recent ground pattern phenomena in the Rhinog mountains, North Wales. *Geogr. Ann.*, Ser. A, Vol. 51, 121-126.

Griffiths, D. H. and Gibb, R. A. 1965. Bouguer gravity anomalies in Wales. *Geol. J.*, Vol. 4, 335-341.

— King, R. F. and Wilson, C. D. V. 1961. Geophysical investigations in Tremadoc Bay, North Wales. *Q. J. Geol. Soc. London*, Vol. 117, 171-191.

Hall, G. W. 1975. *The gold mines of Merioneth.* (Gloucester: Griffin Publications.) 120 pp.

Harms, J. C. and Fahnestock, R. K. 1965. Stratification, bed forms and flow phenomena (with an example from the Rio Grande). *Pub. Soc. Econ. Palaeont. Miner., Tulsa*, 84-115.

Harrison, R. K. 1971. The petrology of the Upper Triassic rocks in the Llanbedr (Mochras Farm) Borehole. *Pp. 37-72 in* The Llanbedr (Mochras Farm) Borehole. Woodland, A. W. (Editor.) *Inst. Geol. Sci. Rep.*, No. 71/18.

Hatch, F. H., Wells, A. K. and Wells, M. K. 1956. *The petrology of the igneous rocks.* (London: Thomas Murby & Co.) 469 pp.

Haynes, J. R. 1981. *Foraminifera.* (London: Macmillan.) 433 pp.

— Kiteley, R. J., Whatley, R. C. and Wilks, P. J. 1977. Microfaunas, microfloras and the environmental stratigraphy of the Late Glacial and Holocene in Cardigan Bay. *Geol. J.*, Vol. 12.

Helm, D. G., Roberts, B. and Simpson, A. 1963. Polyphase folding in the Caledonides south of the Scottish Highlands. *Nature, London*, Vol. 200, 1060-1062.

Henningsmoen, G. 1957. The trilobite family Olenidae. *Skr. Norske Vidensk.-Akad. 1, Mat.-Nat. K1.*, No. 1, 303 pp.

Henwood, W. J. 1856. Notice of the Copper Turf of Merioneth. *Rep. R. Instn. Cornwall*, 41-43.

Herbert-Smith, M. 1971a. The palynology of the Mochras Borehole. Unpublished Ph.D thesis, University of Wales, Aberystwyth.

— 1971b. Palynology of the Tertiary and Pleistocene deposits of the Llanbedr (Mochras Farm) Borehole. *Pp. 93-101 in* The Llandbedr (Mochras Farm) Borehole. Woodland, A. W. (Editor). *Rep. Inst. Geol. Sci.*, No. 71/18.

— 1979. The age of the Tertiary deposits of the Llanbedr (Mochras Farm) Borehole as determined from palynological studies. *Rep. Inst. Geol. Sci.*, No. 78/24, 15-29.

Hofmann, H. J. 1975. Bolopora not a bryozoan, but an Ordovician phosphatic oncolitic concretion. *Geol. Mag.*, Vol. 112, 523-526.

Hsu, K. J. 1964. Cross-laminations in graded bed sequences. *J. Sediment. Petrol.*, Vol. 34, 379-388.

Huddart, L. H. L. 1904. St David's gold mine, North Wales. *Trans. Instn. Min. Metall.*, Vol. 14, 199-219.

Hughes, C. J. 1972. Spilites, keratophyres, and the igneous spectrum. *Geol. Mag.*, Vol. 109, 513-527.

Hunt, R. 1875. The gold mines of North Wales. *Dr Ure's Dictionary of Arts, Manufactures and Mines*, 7th Edition, Vol. 2, 689-698.

— 1881. Mineral statistics of the United Kingdom of Great Britain and Ireland. *Mem. Geol. Surv. G.B.*, 1855-1881.

— 1887. *British Mining*, 2nd Edition. (London: Crosby Lockwood.) 244 pp.

Ineson, P. R. and Mitchell, J. G. 1975. K-Ar isotopic age determinations from some Welsh mineral localities. *Trans. Instn. Min. Metall.*, Sect. B., Vol. 84, 7-16.

Irvine, T. N. and Baragar, W. R. A. 1971. A guide to the chemical classification of the common volcanic rocks. *Can. J. Earth Sci.*, Vol. 8, 523-548.

IVIMEY-COOK, H. C. 1971. Stratigraphical palaeontology of the Lower Jurassic of the Llanbedr (Mochras Farm) Borehole. *Pp. 87–92 in* The Llanbedr (Mochras Farm) Borehole. WOODLAND, A. W. (Editor). *Rep. Inst. Geol. Sci.*, No. 71/18.

JEHU, R. M. 1926. The geology of the district around Towyn and Abergynolwyn (Merioneth). *Q. J. Geol. Soc. London*, Vol. 82, 465–489.

JENNINGS, A. V. and WILLIAMS, G. J. 1891. Manod and the Moelwyns. *Q. J. Geol. Soc. London*, Vol. 47, 368–383.

JOHNSON, B. 1975. Upper Domerian and Toarcian foraminifera from the Llanbedr (Mochras Farm) Borehole, North Wales. Unpublished Ph.D thesis, University of Wales, Aberystwyth.

— 1977. Ecological ranges of selected Toarcian and Domerian (Jurassic) foraminiferal species from Wales. *Pp. 545–556 in* First International Symposium on Benthonic foraminifera of Continental Margins. Part B. SCHAFER, C. T. and PELLETIER, B. R. (Editors). *Maritime Sediments, Spec. Pub.*, No. 1.

KINGSBURY, A. W. G. 1965. Tellurbismuth and meneghinite, two minerals new to Britain. *Mineral. Mag.*, Vol. 35, 424–425.

KOLELAAR, B. P. 1977. The igneous history of the Rhobell Fawr area, Merionethshire, North Wales. Unpublished Ph.D thesis, University of Wales, Aberystwyth.

— 1979. Tremadoc to Llanvirn volcanism on the south east side of the Harlech dome (Rhobell Fawr), North Wales. Pp. 591–596 in *The Caledonides of the British Isles – reviewed.* HARRIS, A. L., HOLLAND, C. H. and LEAKE, B. E. (Editors). *Spec. Pub. Geol. Soc. London*, No. 8.

— FITCH, F. J. and HOOKER, P. J. 1982. A new K-Ar age from the uppermost Tremadoc of North Wales. *Geol. Mag.*, Vol. 119, No. 2, 207–211.

KOPSTEIN, F. P. H. W. 1954. Graded bedding of the Harlech Dome. *Diss., Groningen Rijksuniv*, 1–97.

KRUMBEIN, W. C. and SLOSS, L. L. 1963. *Stratigraphy and Sedimentation.* 2nd edition. (San Francisco and London: W. H. Freeman and Company.) 660 pp.

KUENEN, PH. H. 1953. Graded bedding, with observations on Lower Palaeozoic rocks of Britain. *Verh. K. Nederland. Akad. Wet.*, Afd. Nat (B), 20 (3).

LAKE, P. 1906–46. British Cambrian trilobites. *Monogr. Palaeontographical Soc. London*, Part 1 (1906) 1–28; Part 3 (1908) 49–64; Part 4 (1913) 65–88; Part 5 (1919) 89–120; Part 9 (1935) 197–224; Part 14 (1946) 333–350.

LEWIS, H. P. 1926. On *Bolopora undosa* gen. et. sp. nov.: a rock-building bryozoan with phosphatized skeleton, from the basal Arenig rocks of Ffestiniog (North Wales). [With a stratigraphical note by Prof. W. G. Fearnsides.]. *Q. J. Geol. Soc. London*, Vol. 82, 411–427.

— 1936. Ordovician succession at S.W. end of Aran range, Merionethshire. *Rep. Br. Assoc. Adv. Sci. for 1936 (Blackpool)*, 351–352.

LEWIS, W. H. 1967. *Lead Mining in Wales.* (Cardiff: University of Wales Press.) 415 pp.

LOCKLEY, M. G. and WILCOX, C. J. 1979. A Lower Cambrian brachiopod from the Harlech Dome. *Geol. Mag.*, Vol. 116, 63–64.

LORD, A. R. 1978. The Jurassic. Part 1 (Hettangian-Toarcian). Pp. 189–212 in *A stratigraphical index of British Ostracoda.* BATE, R. H. and ROBINSON, E. (Editors). *Geol. J. Spec. Issue,* No. 8. (Liverpool: Seel House Press.)

LOWE, S. 1981. Radiocarbon dating and stratigraphic resolution in Welsh late glacial chronology. *Nature, London*, Vol. 293, 210–212.

LU YANHAO, ZHOU ZHIYI and ZHOU ZHIQIANG. 1981. Cambrian-Ordovician boundary and their related trilobites in the Hangula region, W. Nei Monggol. *Bull Xi'an Inst. Geol. Miner. Resour, (Chinese Acad. Geol. Sci.),* Vol. 2, No. 1, 1–22.

LYNAS, B. D. T. 1970. Clarification of the polyphase deformations of North Wales Palaeozoic rocks. *Geol. Mag.*, Vol. 107, 505–510.

— 1973. The Cambrian and Ordovician rocks of the Migneint area. *J. Geol. Soc. London*, Vol. 129, 481–503.

MASSON SMITH, D. J. 1971. Geophysical investigations in the Llanbedr (Mochras Farm) Borehole. *In* WOODLAND, A. W., 1971, *q.v.*

MATLEY, C. A. and WILSON, T. S. 1946. The Harlech Dome, north of the Barmouth estuary. *Q. J. Geol. Soc. London*, Vol. 102, 1–40.

MITCHELL, G. F. 1972. The Pleistocene history of the Irish Sea: second approximation. *Sci. Proc. R. Dublin Soc.*, Vol. 10, 187–199.

MOHR, P. A. 1956. A geochemical study of the lower Cambrian manganese ore of the Harlech Dome, North Wales. *Symposium sobre yacimientos de manganese*, XX, I.G.C. (Mexico), Part V, 273–289.

— 1959. A geochemical study of the shales of the Lower Cambrian manganese shale group of the Harlech Dome, North Wales. *Geochim. cosmochim. Acta*, Vol. 17, 186–200.

— 1964. Genesis of the Cambrian manganese carbonate rocks of North Wales. *J. Sediment. Petrol.*, Vol. 34, 819–829.

— and ALLEN, R. 1965. Further considerations on the deposition of the Middle Cambrian manganese carbonate beds of Wales and Newfoundland. *Geol. Mag.*, Vol. 102, 328–337.

MOORBATH, S. 1962. Lead isotope abundances in mineral occurrences in the British Isles and their geological significance. *Philos. Trans. R. Soc.*, Vol. 254A, 295–360.

MORGAN-REES, D. 1969. *Mines, Mills and Furnaces.* (London: HMSO.)

MORRISON, T. A. 1975. *Goldmining in Western Merioneth.* (Llandysul: Merioneth Historical and Records Society.) 98 pp.

NASH, J. T. 1976. Fluid-inclusion petrology—Data from porphyry copper deposits and applications to exploration. *Prof. Pap. U.S. Geol. Surv.*, No. 907-D, 1–16.

NICHOLAS, T. C. 1915. The geology of the St Tudwal's Peninsula (Carnarvonshire). *Q. J. Geol. Soc. London*, Vol. 71, 83–143.

O'SULLIVAN, K. N. 1971. The sedimentology of the Tertiary and Pleistocene beds of the Mochras Borehole. Unpublished Ph.D thesis, University of Wales, Aberystwyth.

— 1979. The sedimentology, geochemistry and conditions of deposition of the Tertiary rocks of the Llanbedr (Mochras Farm) Borehole. *Rep. Inst. Geol. Sci.*, No. 78/24, 1–13.

— IVIMEY-COOK, H. C., LEWIS, B. J. and HARRISON, R. K. 1971. The log of the Llanbedr (Mochras Farm) Borehole. *Pp. 11–36 in* The Llanbedr (Mochras Farm) Borehole. WOODLAND, A. W. (Editor). *Rep. Inst. Geol. Sci.*, No. 71/18.

PEARCE, J. A. and CANN, J. R. 1973. Tectonic setting of basic volcanic rocks determined using trace element analyses. *Earth Planet. Sci. Lett.*, Vol. 19, 290–300.

PETERS, L. J. 1949. The direct approach to magnetic interpretation and its practical application. *Geophysics*, Vol. 14, 290–320.

PETTIJOHN, F. T. 1957. *Sedimentary rocks.* 2nd edition. (New York: Harper and Brothers.) 718 pp.

PHILLIPS, L. 1918. Report of the Comptroller of the Department for the development of mineral resources in the United Kingdom. *Ministry of Munitions of War.* (London: HMSO.)

PHILLIPS, W. E. A., STILLMAN, C. J. and MURPHY, T. 1976. A Caledonian plate tectonic model. *J. Geol. Soc. London*, Vol. 132, 579–609.

Phillips, W. J. 1972. Hydraulic fracturing and mineralisation. *J. Geol. Soc. London*, Vol. 128, 337–359.

Picard, M. D. and High, L. R. Jr. 1973. Sedimentary structures of ephemeral streams. *Developments in Sedimentology*, Vol. 17. (Amsterdam: Elsevier.)

Ponsford, D. R. A. 1955. Radioactivity studies of some British sedimentary rocks. *Bull. Geol. Surv. G.B.*, No. 10, 24–55.

Powell, D. W. 1956. Gravity and magnetic anomalies in North Wales. *Q. J. Geol. Soc. London*, Vol. 111, 375–397.

Price, N. B. 1963. The geochemistry of the Menevian rocks of Wales. Unpublished Ph.D thesis, University of Wales, Aberystwyth.

Ramsay, A. C. 1866. The geology of North Wales. *Mem. Geol. Surv. G.B.*, Vol. 3 (2nd edition), 381 pp.

— 1881. The geology of North Wales. 2nd edition. *Mem. Geol. Surv. G.B.*, Vol. 3, 611 pp.

Rast, N. 1969. The relationship between Ordovician structure and volcanicity in Wales. Pp. 305–336 in *The Pre-Cambrian and Lower Palaeozoic rocks of Wales.* Wood, A. (Editor). (Cardiff: University of Wales Press.)

Readwin, T. A. 1862. On the gold of North Wales. *Rep. Br. Assoc. Adv. Sci. (for 1861, Manchester), Trans.*, Sect. C, 129–130.

Ricci-Lucchi, F. 1975. Depositional cycles in turbidite formations. *J. Sediment. Petrol.*, Vol. 45, 3–43.

Rice, R. and Sharp, G. 1976. Copper mineralisation in the forest of Coed-y-Brenin, Wales. *Trans. Instn. Min. Metall.*, Sect. B, Vol. 85, 1–13.

Ridgway, J. 1971. The stratigraphy and petrology of Ordovician volcanic rocks adjacent to the Bala Fault in Merionethshire. Unpublished Ph.D thesis, University of Liverpool.

— 1975. The stratigraphy of Ordovician volcanic rocks on the southern and eastern flanks of the Harlech Dome in Merionethshire. *Geol. J.*, Vol. 10, 87–106.

— 1976. Ordovician palaeogeography of the southern and eastern flanks of the Harlech Dome, Merionethshire, North Wales. *Geol. J.*, Vol. 11, 121–136.

Roberts, B. 1967. Succession and structure in the Llwyd Mawr Syncline, Caernarvonshire, North Wales. *Geol. J.*, Vol. 5, 369–390.

Rowlands, B. M. 1970. The glaciation of the Arenig Region. Unpublished Ph.D thesis, University of London.

— 1980. Meltwater phenomena in the area between Bala and Corwen, North Wales. *Geol. J.*, Vol. 14, 159–170.

Rushton, A. W. A. 1974. The Cambrian of Wales and England. Pp. 43–121 in *Cambrian of the British Isles, Norden and Spitzbergen. Lower Palaeozoic Rocks of the World.*, Vol. 2. Holland, C. H. (Editor.) (London, New York, Sydney and Toronto: John Wiley & Sons.)

— 1982. The biostratigraphy and correlation of the Merioneth–Tremadoc Series boundary in North Wales. Pp. 41–59 in *The Cambro-Ordovician boundary: sections, fossil distributions and correlations.* Bassett, M. G. and Dean, W. T. (Editors.) Geological Series No. 3. (National Museum of Wales: Cardiff.) 227 pp.

Saunders, G. E. 1968. A fabric analysis of the ground moraine deposits of the Lleyn Peninsula of southwest Caernarvonshire. *Geol. J.*, Vol. 6, 105–118.

Sedgwick, A. 1844. On the older Palaeozoic (Protozoic) rocks of North Wales. *Q. J. Geol. Soc. London*, Vol. 1, 5–22.

— 1852. On the classifiation and nomenclature of the Lower Palaeozoic rocks of England and Wales. *Q. J. Geol. Soc. London*, Vol. 8, 136–168.

— and Murchison, R. I. 1835. On the Silurian and Cambrian systems, exhibiting the order in which the older sedimentary strata succeed each other in England and Wales. *Rep. Br. Assoc. Adv. Sci. (for 1835 Dublin), Trans.*, Sect., 59–61.

Shackleton, R. M. 1952. The structural evolution of North Wales. *Liverpool, Manchester Geol. J.*, Vol. 1, 261–297.

Shepherd, T. J. and Allen, P. M. 1985. Metallogenesis in the Harlech Dome, North Wales: A fluid inclusion interpretation. *Mineral. Deposita*, Vol. 20, 159–168.

Smith, A. J. 1959. Structures in the stratified late-glacial clays of Windermere, England. *J. Sediment. Petrol.*, Vol. 29, 447–453.

Smith, I. F. and Burley, A. J. 1979. A geophysical study in the Cheshire Basin to investigate its geothermal potential. *Rep. Appl. Geophysics Unit, Inst. Geol. Sci.*, No. 70. (Unpublished.)

— and McCann, D. M. 1978. Geophysical investigations. *In* Allen, P. M. and Jackson, A. A., 1978, *q.v.*

Steiger, R. H. and Jager, E. 1977. Subcommission on Geochronology: Convention on the use of decay constants in geo- and cosmochronology. *Earth Planet. Sci. Lett.*, Vol. 36, 359–362.

Streckeisen, A. 1976. To each plutonic rock its proper name. *Earth Sci. Rev.*, Vol. 12, 1–33.

Synge, F. M. 1977. The coasts of Leinster (Ireland). Pp. 199–222 in *The Quaternary History of the Irish Sea.* Kidson, C. and Tooley, M. J. (Editors.) (Liverpool: Seel House Press.)

Taylor, K. and Rushton, A. W. A. 1972. The pre-Westphalian geology of the Warwickshire Coalfield. *Bull. Geol. Surv. G.B.*, No. 35, 150 pp.

Thomas, T. M. 1961. *The mineral wealth of Wales and its exploitation.* (Edinburgh and London: Oliver & Boyd.) 248 pp.

Walker, G. P. L. 1971. Compound and simple lava flows and flood basalts. *Bull. Volcanology.*, Vol. 35, 579–590.

Walter, M. R. 1972. A hot spring analog for the depositional environment of Pre-Cambrian iron formations of the Lake Superior region. *Econ. Geol.*, Vol. 67, 965–972.

Warrington, G. 1971. Palynology of the Upper Triassic strata in the Llanbedr (Mochras Farm) Borehole. *Pp. 73–86 in* The Llanbedr (Mochras Farm) Borehole. Woodland, A. W. (Editor.) *Rep. Inst. Geol. Sci.*, No. 71/18.

Washburn, A. L. 1973. *Periglacial processes and environments.* (London: Edward Arnold Ltd.)

Weiss, S. A. 1977. *Manganese: The other uses.* (Metal Bulletin Books Ltd.) 360 pp.

Wells, A. K. 1925. The geology of the Rhobell Fawr District (Merioneth). *Q. J. Geol. Soc. London*, Vol. 81, 463–538.

Westergård, A. H. 1944. Borrningar genom Skånes Alunskiffer 1941–42. *Sveriges Geol. Undersökning*, Avh. Ser. C, No. 459, 45 pp.

Whittington, H. B. 1966. Trilobites of the Henllan Ash, Arenig Series, Merioneth. *Bull. Brit. Museum (Nat. Hist.), Geol.*, Vol. 11, 489–505.

Wilkinson, G. C. 1979. A palynological survey of some Tertiary sediments in the western part of the British Isles. Unpublished C.N.A.A. Ph.D thesis, N.E. London Polytechnic.

— Bazley, R. A. B. and Boulter, M. C. 1980. The geology and palynology of the Oligocene Lough Neagh clays of Northern Ireland. *J. Geol. Soc. London*, Vol. 137, 65–75.

— and Boulter, M. C. 1980. Oligocene pollen and spores from the western part of the British Isles. *Palaeontographica*, Abt. B., Vol. 175, 27–83.

Wolfe, J. A. 1980. Fluidization versus phreatomagmatic explosions in breccia pipes. *Econ. Geol.*, Vol. 75, 1105–1109.

WOODLAND, A. W. 1938. Petrological studies in the Harlech Grit Series of Merionethshire. *Geol. Mag.*, Vol. 75, 366–382, 440–454.

— 1939. The petrography and petrology of the Lower Cambrian manganese ore of West Merionethshire. *Q. J. Geol. Soc. London*, Vol. 95, 1–35.

— (Editor.) 1971. The Llanbedr (Mochras Farm) Borehole. *Rep. Inst. Geol. Sci.*, No. 71/18, 115 pp.

ZALASIEWICZ, J. A. 1981. Stratigraphy and palaeontology of the Arenig area, North Wales. Unpublished Ph.D thesis, University of Cambridge.

— 1984. A re-examination of the type Arenig Series. *Geol. J.*, Vol. 19, 105–124.

— *In press.* Graptolites from the type Arenig Series. *Geol. Mag.*

APPENDIX 1

Boreholes and other data held on Sheet 135

Boreholes

Two major boreholes have been drilled in the district: these are the Mochras Farm Borehole, details of which have been published by Woodland (1971), and the Bryn-teg Borehole, written up by Allen and Jackson (1978).

During exploration for the Coed y Brenin porphyry copper in 1968–72, 110 boreholes, totalling over 14 000 m were drilled by Riofinex Ltd in the area around Capel Hermon. The core is kept in the BGS rock store. Copies of the borehole logs are held in BGS offices.

A large number of shallow boreholes were drilled into solid ànd superficial deposits during major road improvement schemes at Llanelltyd and Pont Dolgefeiliau on the Trawsfynydd–Llanelltyd road (A487) and on the Dolgellau by-pass. Site investigation reports on these schemes are held in the BGS office at Aberystwyth.

Geophysical data

Aeromagnetic coverage of Sheet 135, obtained during the Aeromagnetic Survey of Great Britain (Geological Survey, 1965) is available from the Regional Geophysics Research Group of the Geological Survey as plots of contours and flight lines at the scale of 1:63 360. The original analogue flight records may be inspected.

A low-level airborne geophysical survey was carried out during 1972–73 over the eastern and south-eastern parts of Sheet 135. Magnetic, electromagnetic and radiometric measurements were taken. The data are plotted on 1:10 560 scale maps, copies of which may be obtained from the Geological Survey office at Keyworth.

Gravity data covering the district are obtainable from the Regional Geophysics Research Group, Keyworth, where the regional data are held as card-image files on magnetic tape.

Geochemical data

As part of the Mineral Reconnaissance Programme the whole of Sheet 135 was covered by a geochemical drainage survey at a density of about 1 site per square kilometre. Water, sediment and panned concentrate samples were taken from each site and analysed for a range of elements. The results are held by the Applied Geochemistry, Applied Mineralogy and Metallogenesis Research Group of the Geological Survey (see also Allen and others, 1979; Cooper and others, 1983).

Photographs

A collection of colour photographs, slides, and black and white pictures of geological features in the Harlech district is held in various Geological Survey libraries.

Palaeontological collections

A large number of fossils was collected from many sites in the district. Many of the important localities are given in papers by Allen, Jackson and Rushton (1981) and Rushton (1982). All the fossils are in the Biostratigraphy Research Group collections, and the details of all the localities and the fossils determinated from them are kept in their records.

APPENDIX 2

Physical properties of rocks

In order either to establish the appropriate geophysical technique for a particular geological problem or to identify the cause of an anomaly it is necessary to know values for the physical properties of the rocks likely to be encountered. Selected values of saturated density, P-wave seismic velocity and magnetic susceptibility have been extracted from published surveys in the Harlech district, and are compiled in Table 14. Further details may be contained in the source references which are indicated by the numbers in parentheses in the table. The method employed to establish the value is also given, since this may have bearing on the validity of the data.

Within the Palaeozoic and Precambrian rocks there is a limited range of densities, with the slate- and mudstone-dominated formations having densities grouped around 2.77 Mg m^{-3}, and the greywacke-dominated formations grouping around 2.73 Mg m^{-3}. The implication is that only relatively small Bouguer gravity anomalies are likely over the outcrop of these rocks, unless bodies of greatly differing composition are present at depth. Rocks younger than Palaeozoic, however, show a great range of density and, where present, would cause substantial anomaly variation.

There is a large scatter of seismic velocities. Pleistocene and Tertiary rocks predictably have very low values. The Jurassic sample may not be representative, but the measured value is surprisingly high. There is significant variation within the indurated Cambrian rocks, but a systematic relationship with lithology cannot be recognised. It is clear that refraction and reflection events could be expected from several horizons, but that velocities alone cannot be diagnostic of rock type.

Susceptibilities are variable both between and within individual formations, with sedimentary and intrusive rocks ranging from effectively non-magnetic to highly magnetic. Consequently, strong magnetic anomalies might be expected.

A series of laboratory determinations of resistivity by Forster (1976) have been analysed in Allen, Cooper and Smith (1979) for the cause of electromagnetic anomalies. The values show variation between 1 and 3400 Ω m for saturated samples of the Dolgellau Member of the Cwmhesgen Formation and for the Clogau Formation. Unpublished investigations of similar samples using a resistance meter have demonstrated that certain current paths along sulphide-filled fractures have effectively zero resistance. Thus specific strata within these formations have very low resistivity indeed, and could be detected using electrical geophysical prospecting methods.

Table 14 Selected physical properties of rocks from the Harlech district

Unit (F. = Formation) (G. = Group)	Saturated density Mg/m^3	P-wave velocity km/s	Susceptibility $\times 10^{-3}$ SI units
Pleistocene	2.5[1,a] 1.95[9,p] 2.25[1,b]	1.5[9,p]	
Tertiary	2.25[1,c]	2.41[1,c] 2.19[9,c]	
Jurassic	2.57[1,h]	4.26[1,h]	
Triassic		3.08[6,p]	
Aran Volcanic G.	2.65–2.70[2,d]		25[13,m]
Rhobell Volcanic G.	2.75[2,e]		0.2[13,n]
Cwmhesgen F.	2.77[2,f] 2.68[10,f]		
Ffestiniog Flags F.		4.86–5.49[4,f] 5.43[8,f]	
Maentwrog F.	2.79[10,j]	6.1[4,i] 5.49[8,i]	3[13,o]
Upper Cambrian	2.73[3,p]		
Clogau F.	2.77[10,f]	5.67[8,f]	0.47–3.49[10,f]
Gamlan F.	c.2.7[9,f]	6.4[4,f] 5.18[7,f] 4.88[8,f]	
Barmouth F.	2.69[2,g] c.2.7[9,g]	5.49[4,g]	
Rhinog F.	2.72[10,g]	3.90–4.97[7,g] 5.15[8,g]	19.8[10,g]
Llanbedr F.	2.85[9,f]	3.96[4,5,f] 3.99[7] 6.31[8,f]	
Dolwen F.	2.73[11,g]	4.57[5] 5.0[7,g] 5.18[8,g]	0.13–125.7[11] 80[13,g]
Cambrian	2.7–2.8[4,p] 2.74[12,p]	5.67[5,l]	
Precambrian	2.77[11,k]	5.46[11,k]	0.54–42.7[11,k]

Notes on Table 14

Numbered annotations refer to source of data.

1 Masson Smith *in* Woodland (1971); density and sonic borehole logs.
2 Powell (1956); laboratory measurements.
3 Powell (1956); gravity traverses.
4 Griffiths, King and Wilson (1961); *in situ* measurements.
5 Griffiths, King and Wilson (1961); laboratory measurements.
6 Hill, M. N. reported in discussion *in* Griffiths, King and Wilson (1961); seismic refraction.
7 Blundell, King and Wilson (1964); field measurements.
8 Blundell, King and Wilson (1964); laboratory measurements.
9 Blundell, Griffiths and King (1969); gravity and seismic profiles.
10 Forster (1976); laboratory measurements.
11 Smith and McCann (1978); borehole log.
12 Griffiths and Gibb (1965); combination.
13 Allen, Cooper and Smith (1979); field measurements.

Lettered annotations refer to rock type as described by authors.

a Boulder clay.
b Clays and silts.
c Silts, sands and clays.
d Grits and ashes.
e Intermediate lavas.
f Slates.
g Grits.
h Incomplete section.
i Vigra Flags member.
j Penrhos Shale member.
k Bryn-teg Volcanic Formation.
l Mean for Cambrian refractor.
m Dolerite.
n Tuff.
o Minor intrusions in Maentwrog Formation.
p Not specified.

INDEX OF FOSSILS

Note: In compiling this list, signs which qualify identification (e.g. cf., aff., ? etc.) have been ignored.

GENERAL INDEX

Note: Page numbers in italics refer to illustrations

Kames 94
Kettle holes 97

HER MAJESTY'S STATIONERY OFFICE

HMSO publications are available from:

HMSO Publications Centre
(Mail and telephone orders)
PO Box 276, London SW8 5DT
Telephone orders (01) 622 3316
General enquiries (01) 211 5656

HMSO Bookshops
49 High Holborn, London WC1V 6HB
(01) 211 5656 (Counter service only)
258 Broad Street, Birmingham B1 2HE (021) 643 3757
Southey House, 33 Wine Street, Bristol BS1 2BQ
(0272) 24306/24307
9 Princess Street, Manchester M60 8AS (061) 834 7201
80 Chichester Street, Belfast BT1 4JY (0232) 234488
13a Castle Street, Edinburgh EH2 3AR (031) 225 6333

HMSO's Accredited Agents
(see Yellow Pages)

And through good booksellers

BRITISH GEOLOGICAL SURVEY

Keyworth, Nottinghamshire NG12 5GG
Murchison House, West Mains Road, Edinburgh EH9 3LA

The full range of Survey publications is available through the Sales Desks at Keyworth and Murchison House. Selected items are stocked by the Geological Museum Bookshop, Exhibition Road, London SW7 2DE; all other items may be obtained through the BGS Information Point in the Geological Museum. All the books are listed in HMSO's Sectional List 45. Maps are listed in the BGS Map Catalogue and Ordnance Survey's Trade Catalogue. They can be bought from Ordnance Survey Agents as well as from BGS.

On 1 January 1984 the Institute of Geological Sciences was renamed the British Geological Survey. It continues to carry out the geological survey of Great Britain and Northern Ireland (the latter as an agency service for the government of Northern Ireland), and of the surrounding continental shelf, as well as its basic research projects. It also undertakes programmes of British technical aid in geology in developing countries as arranged by the Overseas Development Administration.

The British Geological Survey is a component body of the Natural Environment Research Council.

Printed for Her Majesty's Stationery Office by Linneys Colour Print Ltd.
Dd 737389 C20 1/86 46009